Warm Greenhouse Plants

by the same author

Grow Fruit in your Greenhouse

Warm Greenhouse Plants
and
Their Cultivation

George E. Whitehead

Faber and Faber
London

First published in 1974
by Faber and Faber Limited
3 Queen Square, London W.C.1
Printed and bound by
Butler & Tanner Ltd, Frome and London

ISBN 0 571 10173 9

The Secret influences of daylight and atmosphere, the thousand pangs emanating from all Nature, are the ransom of sensuous beings, prone to find their delight in the shapes and colours of things.

The Garden of Epicurus by Anatole France

Acknowledgements

IT is impossible to compile a book of this sort without help from others. Among those who have given personal assistance are Brian Charman, who did most of the drawings, Ann Bonar, who spent hours helping to arrange the order of the chapters and species, and Pamela Harrington, who so painstakingly deciphered the original script and retyped where necessary.

It is obvious that a reference book of this nature could not be completed without referring to other books. Some are mentioned in the Introduction, but often the information recurs again and again and it is impossible to trace it back to the original source. Also, having worked with other learned gardeners from whom I have imbibed much, I have long realised that knowledge, be it practical or theoretical, comes from so many unexpected beginnings that, much as we would like to, we cannot always give the credit that is due. I am just humbly thankful.

To the following, who supplied photographs, I am very grateful: Mr Ernest Crowson (monochrome plates 2, 3, 9, 10, 12–15, 18, 19), Mr Ron Oulds (colour plate 9), Mr Harry Smith (colour plates 1–6, 10–13 and monochrome plates 1, 4–8, 11, 16, 17), Messrs G. F. Strawson & Son of Horley, Surrey (monochrome plates 20, 21) and Messrs Donald Wright & Associates, Industrial Design Consultants, Durham (on behalf of the manufacturers, Messrs W. Richardson & Company of Faverdale, Darlington) (monochrome plates 22, 23).

George E. Whitehead

Contents

Part Three

Monochrome Illustrations

Colour Illustrations

(*between pages* 136 *and* 137)

Line Drawings

Introduction

WHEN the stately homes of Britain were in their heyday, there were impressive conservatories and ornamental greenhouses sufficiently heated to contain the most exciting tropical plants collected from the equatorial regions of the world. These amenities were popular within the restricted circle of wealthy nobility for, not only did they give a degree of status, they provided facilities and means of pleasure to those guests who preferred to relax amidst the palms, ferns and flowering plants in a congenial atmosphere, as an alternative to the rigours of outdoor sports. It was a wonderful combination of the best of both worlds. Then came the two wars that practically ended that way of living.

But the interest in tropical plants did not decline. The gardeners who cultivated them, after returning from serving in the Forces, found their previous positions redundant and so obtained employment in municipal and public parks and gardens, taking with them not only their skill, but often the beginnings of collections that have aroused real interest among a much larger and widespread section of the populace.

At a recent Chelsea Show, I counted the number of stands in the large marquee where greenhouse or house-plants were featured and was delighted to discover that they numbered no less than a quarter of the total, including all the more supposedly popular subjects such as roses, shrubs and herbaceous flowers. And these did not include fuchsias or geraniums.

The importation of these adorable subjects is an increasingly thriving business. Most come from the Continent, for many affluent Europeans grow them in their homes, just as we cultivate our own house-plants. They are also widely distributed in the United States. Many arrivals are consigned to florists who use them to decorate hotel lounges, theatre vestibules and hospital reception rooms.

These individual plants, like other living things, are either lucky or unlucky. Through neglect, or faulty care, some linger longingly until they perish. They are put in draughty corners, or placed in dry rooms, instead of being kept from draughts, or given the moisture they require. The lucky ones are given homes in greenhouses that are maintained at a right degree of heating, and the atmosphere is controlled to their liking. That is what this book is all about.

I am trying to show that the care of tropical plants in a suitable greenhouse is the easiest form of intimate gardening. Regarded in the right way, it is no more difficult than running a home. If it is done in the same spirit as managing a household it gives endless pleasure, for there is constant change through the seasons of the flowers, and their marvellous fragrance, as well as their wonderful forms and colour hues of leafage, of which one never tires.

Some are already popular house-plants. But most of the desirable ones do not like dry atmospheres. However, if one has a greenhouse it becomes a floral decorator's dream, for the inmates can be put into living-rooms, and by constant changes, will not suffer from a week or two's stay in these less congenial surroundings.

I am writing much of this from personal experience. My first contact with these exciting subjects was when being instructed by the head gardener of a large establishment on the methods of watering and syringing the inmates of the conservatory and the warm greenhouses where I then worked, half a century ago. It was nice being in the warmth when the outdoor conditions were bleak. It is difficult to describe my first feelings, although I still remember them well, but it is no exaggeration to say that I was enchanted. As the years passed I went from large garden to large garden, eventually becoming a head gardener myself, where there were considerable staffs in the various departments under my control. But, always, the heated greenhouse, where the tropical plants and the orchids lived quite happily, gave me a special feeling of elation, unequalled in any other part of any garden.

Of course, there is the economic aspect. You must have a greenhouse. You must fit it with appliances and apparatus in order to keep it warm. These factors are considered in more detail as they arise. Here I contend that the outlay is not beyond the means of hundreds of thousands of people, if they would really like to make such a beautiful investment. A complete new

Aphelandra squarrosa louisae. Leaves shiny green, striped white. Flowers yellow.

Begonia rex 'Merry Christmas'. One of many cultivars of a species grown for its unique foliage.

greenhouse with all modern equipment costs less than a luxury car, or swimpool, or hard tennis court.

In fact, people do buy greenhouses. There are thousands manufactured every year. So often, they disappoint. They are just pipedreams that end up as spiders' refuges. If only that little more was spent, and proper installation put in, the outlay would then be a lovely inspiration instead of a failure. The desire was there. By doing the job properly it could have, and can be, fulfilled to the utmost.

The main section of the book comprises a catalogue of suitable plants, together with specific information regarding their cultivation. You may wonder how, should you want them, so many can be obtained. Some have been grown in this country for many years and some are comparatively new.

The best way to start is to make a habit of looking into florists' shop windows. If a plant looks good, and the price is not too prohibitive, you will be tempted to buy. Furthermore, you will soon be able to compare one plant with another and acquire a passable knowledge about them. Another source is exhibitions. Stands of indoor plants are seen increasingly at large exhibitions and County Agricultural events. If you boldly buttonhole the attendants they will, most times, be pleased to chat with you about their living wares.

Another source of supply is the seed catalogue. One sure way to make a start is to get packets of seeds of plants like aechmeas, anthuriums, begonias, crossandras, gloxinias, peperomias, saintpaulias, and so on, and fill the greenhouse cheaply and very quickly. Then as the real treasures are acquired, the unwanted inferior plants can be given away.

Collecting warm greenhouse plants is a pastime much like an antique dealer's livelihood. Sometimes you get a real bargain. Sometimes you get a wrong one. But the difference is that, if you have a good greenhouse and the knowledge, you can propagate from your rare plants, and often make a profit that way, if you are so inclined.

If you become sufficiently keen on any particular family of plants, it is possible that you can contact others with similar tastes through a specialist society. For example, the British Bromeliad Society publishes a quarterly journal and has meetings, no doubt, where the members show and exchange their treasures. That and other organisations are most often affiliated to the Royal Horticultural Society, Vincent Square, London, S.W.1, and it is possible to elicit helpful information from that source by judicious enquiry.

When the comprehensive lists of plants are looked through by experts they may criticize my selections. They will wonder why many have been included and others of superior quality have been left out. Actually, it is not a selection based purely upon quality, but an attempt to make the lists most representative. If you know dipladenias, for example, of which I have mentioned three species, you will be aware that there are scores of others. But one dipladenia is only another dipladenia as far as treatment is concerned—all need much the same. What I have done is to briefly describe the ones I favour, and make you aware of their desirability, as well as point out that there are many others. In smaller genera there can be but one good species that we must know how to cultivate, therefore I have written about it in order to give the essential information.

If I included all the tropical genera and species, and varieties in cultivation, this work would become a mighty tome. I have perforce to be content with a mere outline of the subject.

Many greenhouse plants not mentioned in this book are to be found in *The Cool Greenhouse Today*, by Deenagh Goold-Adams. I purposely refrained from checking my collection of plant names with those in that work until I had completed this one; what surprised me then was how few had been duplicated. Some of those intended for the cool greenhouse will also grow well in a warmer one, but the majority I have named will suffer in the lower temperatures.

Anyone who is interested in, and requires more detailed information on, automation and thermostatic controls than I have given is recommended to consult *The Cool Greenhouse Today*, for it contains much that I have not included, for the reason that I have not given the problem adequate intensive study. What I have done is to indicate the possibilities, but at the same time, keep the whole subject in proper perspective. However, there are continuous developments that one cannot foresee, and the only way to learn about them is to remain ever alert.

While I am all for simplification, whether it be in heating, watering, keeping down pests, ventilation or shading, I still think that personal attention counts most in the end. In all these matters there must be compromise to attain complete success.

One of the headaches is the correct naming of plants. It is most important, for I want to be sure that the plant I am referring to is that one and no other or, if someone else is talking about a plant, that I am not creating confusion.

The character of tropical plants does not change. They are now as they were a thousand years ago, and will be the same a thousand years hence. But the names change. They are given by nomenclaturists rather than gardeners, who find some new characteristic that causes them to be put in different genera. Sometimes they discover that some plant was known by another title in the dim past, so they insist we change to that. In some instances the names have been altered several times. And to this day we can so easily be misled by the label.

Nevertheless, as far as possible I have tried to be correct. I have checked up with authoritative works, such as the Royal Horticultural Society's *Dictionary of Gardening*, published in 1951. I am unaware of any more recent work upon this particular subject. There are several books about house-plants that only cover a tiny part of the subject.

My old reference books, which I studied while young, were Nicholson's *Dictionary of Gardening*, and *Greenhouse and Stove Plants* by T. Baines, published before I was born. Looking through the pages yet again, I have been reminded of facts that I had long forgotten, but which are as applicable today as they were then.

Of course, we now know more about some things. Mr Baines regarded *Choisya ternata* as a warm greenhouse plant. It is perfectly hardy. We have discovered that many things the old gardeners coddled grow better outdoors. But there are many things we have forgotten. As a lifelong student, I appreciate the vast amount of recorded and unrecorded experiments that have been expended upon bringing the cultivation of these wonderful plants to the high standard possible today.

I have written this book because I felt that it could fulfil a useful uptrending purpose. I hope that it will.

Part One

CHAPTER 1

Principles of Management

THE very best way of obtaining the utmost enjoyment from the luxury greenhouse is to embark upon a definite policy of growing only those subjects that will live harmoniously with one another in the conditions provided. In this book we are exclusively concerned with the cultivation of temperate or tropical plants that thrive in either a warm greenhouse, or a warmer one, normally referred to by gardeners as 'the stove'.

The terms 'warm greenhouse' and 'stove' recur many times in the following pages, to distinguish two levels of artificial heat supplied whenever the outdoor weather is below the temperatures we need inside.

Temperatures

A warm greenhouse needs a normal temperature of 16°C (60°F) during the shortest days of winter, with a minimum of 13°C (55°F) in the coldest of cold spells. There will, of course, be much higher levels in spring and summer, but they need not worry us for the moment.

The stove most often has a normal winter temperature of 21°C (70°F), with a minimum of 16°C (60°F) when the outside weather is bitter.

Both centrigrade and Fahrenheit recordings are given. Most thermometers, with maximum and minimum registers, have them. I personally, being familiar with Fahrenheit, find it difficult to combine both, so forget centigrade. The chief thing is to memorise the above temperatures. When the stove, or the warm house, is referred to, there is then no need to include these figures every time.

When you consult the long, but by no means complete, lists of plants in the main section of the book, you will see that some will grow in the warm greenhouse. These will, unless specially stated, likewise flourish in the stove. But those that demand stove conditions are less likely to grow as well in the warm greenhouse. Therefore, if a stove temperature can be maintained, there is a wider choice. Notwithstanding, very large greenhouses can be filled with those that live well in the lower temperature, without duplication of species or varieties.

No doubt the warmness or suitability of the greenhouse will be determined by the heating arrangements. Some houses are so exposed to the vagaries of the weather that it is impossible to keep them warm enough. In a later chapter we will consider greenhouses and their heating in more detail, but at this juncture, I merely give my preference for the type of structure that will retain the maximum of warmth with the minimum of artificial heat.

It should be a fairly large lean-to with solid walls up to 2½ ft. high, while the framework above must be of wood, metal, or aluminium, the roof being neither too steep nor too flat, and fitted with ample ventilators. It should be situated upon the south or south-west side of the residence or a high wall, where there is much sunshine and light, but where it is protected from the coldness of the north and east.

The heating apparatus should keep the temperature evenly distributed throughout the greenhouse, and for this there is nothing better than 4-in. pipes filled with water that is kept hot and made to circulate by a boiler fired either by means of solid fuels such as coal or anthracite, or by domestic fuel oil. If there is oil firing controlled by thermostat, and it is automatically fed, that is the perfect arrangement.

There are several heating systems, including the burning of gas or paraffin, and also the use of electricity. While electricity is excellent for cooler temperatures, it is generally too costly for the warm greenhouse or stove, but it is my third choice after solid fuel or oil. It can be automatically controlled, but can lead to disaster during power cuts, that ironically happen when heat is needed most. There are also heating installations combining electricity with oil (see chapter 25).

Humidity

Practically all tropical plants need an atmosphere that is buoyantly humid. It is necessary to provide it, during spring and summer particularly, by constantly applying water in various ways. Giving water to the plant roots helps but is not enough. The leathery-leaved species—as most are, especially the shrubby evergreens—appreciate being syringed two or three times daily in bright weather. Another and quite usual procedure is 'damping-down', which simply means splashing water about the footways, and on the staging.

Rain collected off the greenhouse roof and retained in a tank inside gives the best supply. There are also mains, but soft water is always preferable. Whatever is used, it is best warmed by being kept in the greenhouse several hours before use.

There are many modern types of gadgets that automatically provide mistified particles of moisture, and these should be carefully considered, for they can be invaluable to those who cannot attend to their plants regularly. If the water comes from the mains, it may lower the temperature while being constantly released. One of the most favoured automatic methods is the one whereby water in the greenhouse is projected into the atmosphere by electrical dispensation.

Although I do not advocate electricity for heating, I do consider it most valuable for creating humidity, and at the same time giving air circulation.

If you walk into a stove or warm greenhouse that is properly heated and the air circulated, you will feel the comfortable, moist atmosphere, but if you enter a cold house you will immediately sense the clamminess. Gardeners often refer to buoyancy. It is actually the amount of water vapour suspended in the air on account of its warmth. If there is also something to agitate the air so that it gently moves about, much the same sort of conditions are provided that plants get in their original homes.

Water

To facilitate watering there are several ways of supplying it automatically through the bottom of the pots, when they are stood on damp or trickle beds that are kept in that state by automatic controls. These aids are very useful, and for those who prefer to use gadgets rather than to care for the plants themselves, they are far better than careless attention. You can then leave the plants for a week or a fortnight, resting assured they will still be alive and fresh when you inspect them. I personally would rather have such devices in reserve for rare occasions such as holidays, and would never rely upon them entirely. In such a case the plants could be stood closely together on the shelving or trays during the period of absence, and then returned to their proper places afterwards.

Staging

My preference is for wooden slatted staging, painted white, and tiered so that each plant gets the maximum of room, and can be seen in all its glory when placed at the most suitable height.

The plant house should be a thing of aesthetic beauty, rather than a utilitarian, labour-saving object. The background should enhance the beauty of the subjects rather than detract from them, as most labour-saving gadgets do. Therefore, I say, have such gadgets for emergencies or occasional convenience, but put them away when personal attention is the alternative. The plants then feel good, and their surroundings are better.

The floor should be of soil, fine shingle, or ashes. In houses of considerable size, the climbing plants that enhance the beauty of the interior will grow in beds if they are well drained. They like to spread their roots, and feel free.

Pathways

The pathways need not be very wide, but should have curbs to contain loose shingle, or whatever other material is used, to a fair depth. During summer it will then retain moisture for a few hours, while in winter, when water is less necessary, the shingle need not be damped but will help to maintain a good even temperature. Concrete has little retentive qualities.

Paths need not be very wide. Nevertheless, a space or two should be left in the general pattern to place seats for visitors or the family, although these must not be left permanently, or they become nests for pests, or are too damp for use.

Another thought is never to have staging along both sides of the path at a similar height. If you do, the chances are that the overhanging plants will be brushed or knocked over by careless walkers. Apart from that, a staggered effect is always more pleasing than a soldierly array. How plants are placed is a matter of taste, but contrasts in heights and character of foliage appeal to me more than uniformity (see Figs 1 and 2).

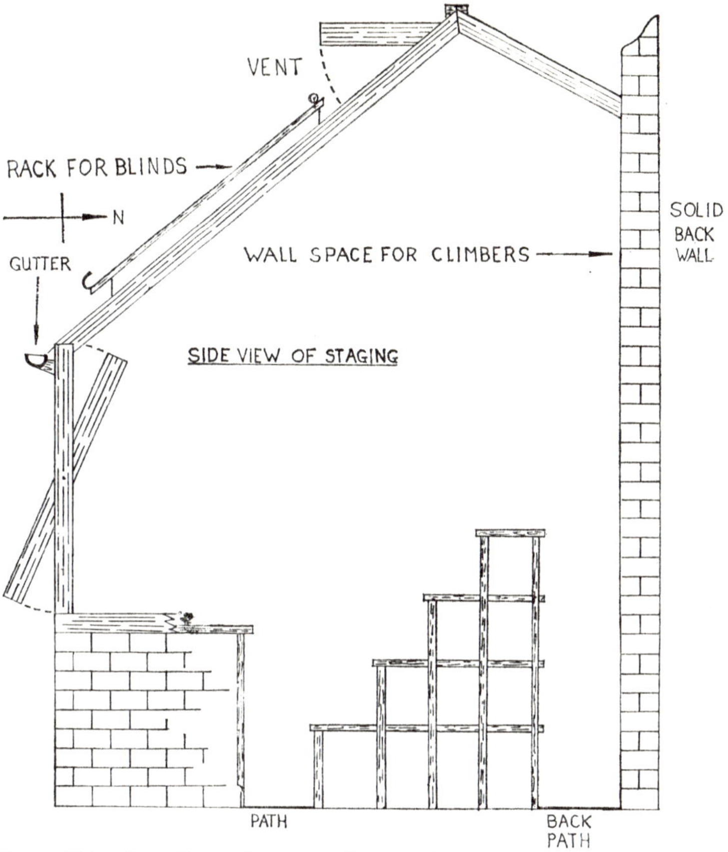

Fig. 1. Side view of greenhouse staging.

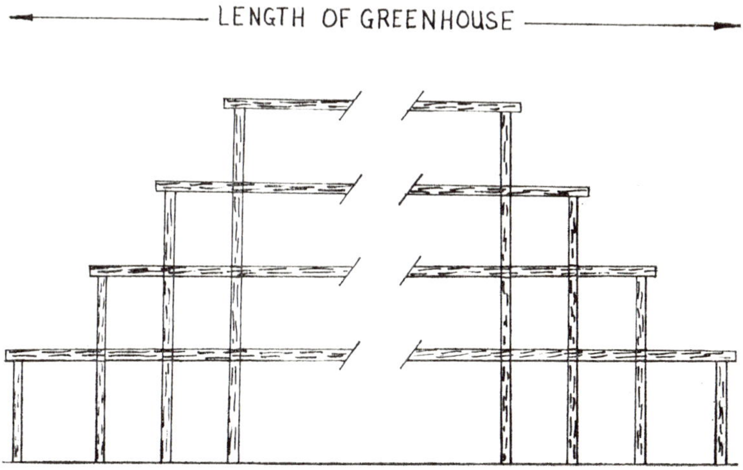

Fig. 2. Frontal view of central staging.

Ventilation

To be able to control temperatures by opening and shutting ventilators during spring and summer is part of the art of being a good grower of plants. That can be done by automation, too. Whether or not it is preferable to doing the opening and shutting personally, as needed, depends upon one's personality. But much can be saved by installing an air circulator. There are available small fan-like devices, worked by electricity, that keep a constant circulation. And if there is a minimum of ventilation given, the moisture in the house is retained.

I am of the opinion that, if a fan can be installed, the ventilators should be kept completely shut from early November until late February. If there is any sunshine it will not boost the heat beyond endurance, and by excluding all air that can enter other than by crevices, such as between panes of glass, much heat is conserved. Furthermore, it may be an advantage to dry out the inside atmosphere a little, rather than have it intermingled with the cold damp prevailing outdoors.

The drawback of thermostatically controlled ventilators is that they cannot anticipate. For example, if the gadget is set to open and shut at 24°C (75°F), it will surely be better if the weather threatens to be hot during the day to start letting in air before that register of warmth is reached. And it most certainly will be advantageous to close all ventilators before the temperature falls below that figure. Yet you must set the automatic opener at one level. Perhaps the best solution is to instal one small ventilator to be worked by automation, as a safety precaution, but carry out main manipulations manually.

Propagation

In addition to the main body of the greenhouse being maintained at a suitable temperature and buoyant atmosphere, there must be two small sections for dealing with the raising of plants, and the resting of others. In the first one a propagator can be installed whose interior is kept warmer and more moist than the main part of the building. It can be a small enclosed glass contraption placed on a bench, heated by electrical cables under its floor, or it can be stood just above hot-water pipes. It is mostly used for cuttings. Sometimes the cuttings are inserted in a sandy bed of soil through which the cables or pipes run.

Most often, if there is a bed of sand just above the warmth suppliers that is kept moist, the cuttings can be inserted in pans of rooting material and placed on that. The glass enclosure holds the moisture in, and so stops the plants from flagging or wilting until they form roots.

That form of propagator is being replaced by an open one with no cover to keep the moisture in. It has a heated floor in the usual manner but, instead of a covered top, has a mist projector that works by an automatic principle. It is in general use in large nurseries where a whole propagating house is filled with cuttings that are kept moist by several mist ejectors working.

If not too difficult to instal, one mist unit placed over the uncovered propagating case will enable the grower to root most subjects from cuttings, or to use whatever other means of increase he may prefer. At the same time, the mist will drift into the rest of the greenhouse and help the plants considerably, as well as reducing the necessity for applying water so often in summer. But it should not be used in the dead of winter (see chapter 25).

Resting

Most evergreens must have some moisture throughout the year, although they need much less in winter than summer. There are deciduous plants that need considerable dryness at certain times of year, but not necessarily at the same time. Therefore, if there is another glass chamber within the house into which they can be put during their resting periods, they are more likely to survive.

The poinsettia is a case in point. After the bracts fall, it needs a baking equivalent to the drought period it has in its native habitat. If popped into the resting chamber it will receive this. This chamber need not be very large, nor ever be near the heating pipes or cables. Placed at one end of the greenhouse, away from the general collection, it will in no way detract from the aesthetic appearance. Also, if there is no potting-shed, a work bench can be included in the same business corner of the house.

Shading

Most tropical plants need much light. But it is wrong to assume that they can stand hot sunshine through glass. During winter they will never get too much. Shading is hardly ever necessary from early September until late April or early May. It is essential during the hot cloudless days of summer. If it is possible to give personal attention, the fitting of adjustable blinds is the best possible method of providing shading. They can be of either cotton material, Netlon or wood laths. My preference is for Netlon of a very small mesh that breaks the sun rays rather than excludes light. Whatever is used, the blinds should be on rollers that can be controlled by cords that work comfortably (see chapter 24).

The alternative is to use one of the patent greenhouse paste sprays that are obtainable in green or white. They are mixed to a creamy consistency with water and brushed on over the glass, as thinly as possible. If the whitewash brush can be deftly wiggled, it results in a stippling effect quite pleasant to look at. Another way to apply the shading is to use a Solo Sprayer that has a longish lance fitted with an adjustable nozzle at the top end, and a suction hose to drop into the container at the other.

Pots

We need pots for our plants. At one time clays were universal. There is still nothing better. There are also paper, whalehide, peat and plastic ones that are of little practical use in the plant house. Unfortunately, not all the plastic pots are as practical as they seem; some lose their lustre and become brittle, and only by trial and error can we ascertain which are good enough for our greenhouses. Then, having found those that are, buy only that brand as long as it is available, for nothing is more marring to a pleasing arrangement than mixed shapes and colours.

There are sizes ranging from 2 in. in diameter up to 15 in. At one time every list contained a number such as 'sixty', which was either 2½ or 3½ in. in diameter. It was when clays were 'cast' in such numbers and was the number of pots of a specified size that could be produced from a given quantity of clay. We now merely give the sizes. Plants need pots that hold the roots happily, and should never be put into any too large for them.

When choosing hard plastic pots, I look to see whether there is a slight rim around the outside of the bottom so that surplus water that escapes through the holes can drain away. If there are plenty of holes, and there is the scalloped edge, there is no need to put crocks in the bottom, as is the case with clay pots, that have only one, larger hole.

There are advantages and disadvantages with both clay and plastic pots. Watering has to be more carefully done when roots are surrounded by plastic material instead of clay. Clay absorbs and gives back moisture. Plastic material is without porosity. But there are both good and bad plastic and clay pots and good plastics are better than bad clays. However, lovely clay pots that stay clean, instead of greening up quickly, serve plants best.

Composts

To the uninitiated, various descriptions of composts such as 'no-soil', 'John Innes', or some other type, may be puzzling. I will try to make the subject as clear as possible.

In days gone by gardeners had their own formulas for various plants, the commonest being:

3 parts garden loam
1 part oak or beech rich leafmould
½ part sharp horticultural sand

The loam was most often the underpart of parkland turf stacked in the potting-shed yard to ripen, as it were, for about a year. The leafmould was two or three years old and had likewise been matured. The sand was bought.

For various plants this formula was changed according to known needs, and experience of cultivation of various families and species of plants. Sometimes horse manure was added, or loam reduced, while quite frequently lime rubble, bonemeal, or a branded fertiliser was added, the usual measure for this being a 4-in. pot to each barrowload of the mixture.

Reference works and dictionaries will often give such formulas now, but several years ago the scientists invented standard potting mixtures to suit almost every need. They are John Innes seed compost and John Innes potting composts. They can be bought from various suppliers, who mix them in large quantities more economically and thoroughly than the gardener can. On the other hand, there are also some of bad quality around, not made up strictly according to the formula, and one must either have complete faith in the firm that deals with these, or make one's own.

The recipe for the John Innes potting compost as frequently mentioned in the section on individual plants is:

7 parts loam
3 parts peat
2 parts coarse or sharp potting sand

The quantities are by loose bulk and not by weight.

The loam has to be sterilised, preferably by steam, in an apparatus where it will penetrate the loam at a temperature of 100°C (212°F) and remain so for about 10 minutes. After it has dried out to the state in which it can be cleanly handled, it is mixed with the peat and sand.

Loam is a mixture of clay and sand, or silica, the amounts present determining whether it is light, medium or heavy. The nature of the loam depends on where it comes from; the best is generally taken off pasture land and stacked to mature before being steamed.

There are two main sorts of peat—sedge and sphagnum moss. Sphagnum moss peat varies according to the district of origin. I prefer Irish (Shamrock) because of its consistency and cleanness. Some of the medium coarse sedge peat from Somerset is quite suitable, when it is of good quality.

The sand can be Welton Grit, Lincoln Grit, Bedford Sand, or some such reputed sort. The chief point is that it should be free from dirt or chemicals.

Having mixed these three ingredients, John Innes base fertiliser is added, consisting of 2 parts superphosphate of lime (18% phosphoric acid), 2 parts hoof and horn meal (13% nitrogen) and 1 part sulphate of potash (48% pure potash, not kainit or muriate). These mixed together are then sprinkled over the already mixed heap at the rate of 4 oz. to each bushel; ¾ oz. of ground chalk is also added to each bushel.

You can buy this John Innes base to add to your own home prepared compost if you like. But you must include your own ground chalk. I find the best policy is to seek for a reliable supplier, be it a national firm or a friendly neighbouring gardener.

In recent years, there have been constantly appearing many brands of no-soil potting composts. They are practically all

Bougainvillea glabra. A moderate climber with wonderful metallic, rose-coloured bracts.

◀ *Calathea louisae*. Closely related to marantas. All with very decorative leaves.

Clivia miniata. The Kaffir lily—stiff, deep evergreen foliage and deep orange flowers. Easy to grow.
▼

based upon either sphagnum or sedge peat, to which is added essential chemical fertilisers that supply the main and trace elements. A few are of doubtful quality, but the chief brands are being improved upon all the time and are attaining a great degree of reliability. Some contain urea, a comparatively new chemical, that supplies nitrogen steadily to the plants.

One of the problems of peat composts is that peat itself contains very little food, and anything added is not retained for any length of time; therefore, further feeding has to be resorted to after a few weeks, whereas John Innes sustains the roots for months.

Another disadvantage I have found with no-soil compost is that the type of plants we are interested in, whose main attractions are the shades of colours in the ornamental leaves, seem to be less bright and less good than when there is loam in the soil.

For many plants named in the text you will read that John Innes potting compost, plus Irish sedge peat, is recommended. If one bushel of peat is added to every three bushels of the John Innes compost, the quantity of peat in the mixture is doubled, yet the longer-lasting and firmer texture given by the loam is retained.

To bring the composting soil problem down to utmost simplicity, there need only be two bought commodities. These are John Innes potting compost No. 1 and Irish moss peat. Alternatively, and reservedly, for ordinary peat there can be a proprietary no-soil compost such as Kerimure, or one of the other reliable brands used as a substitute.

However, for rooting cuttings and leaves as described in the next chapter, the medium must drain properly and yet retain moisture, and for that we get a coarse sharp sand to add to either the John Innes mixture, or to sedge peat. Alternatively, we can use vermiculite, which is a substance that has been specially treated so that it holds moisture and yet is extremely porous. It contains no plant nourishment, but does give moisture to the cuttings and, if mixed with sedge peat, encourages root production.

Fertilisers

Applying fertiliser to the roots of pot plants is an important part of cultivation. Once they become 'potbound', or in the case of the no-soil composts, the nutrient is used up, regular doses of a proprietary plant food are essential. Proprietary brands such as Sangral, Phostrogen or Maxicrop are most suitable. Where there are special formulas for house-plants they should be chosen. But be careful, because excessive feeding is a very bad thing, and can lead to as much trouble as too little.

Pests

A properly heated stove or warm greenhouse with the air kept buoyantly moist should hardly ever be invaded by either insects or fungal pests. However, when they do appear, there must be a means of destroying them before they spread, which they will do rapidly, if allowed.

The most likely intruders are ants and woodlice. Ants are attracted by sweetness. The main damage they do is to carry other pests from plant to plant, chiefly greenfly, that in turn spread virus diseases. While these troubles are most serious outdoors and in cooler greenhouses, they will seldom get out of hand amongst our plants. But in case they do, there are several branded ant-killers on the market.

Some of these ant-killers contain BHC, which is a chemical relative of the notorious DDT, and will also kill woodlice, which can be a nuisance in warm greenhouses. However, the best precaution is to keep plants free from rubbish and woodwork clean.

Woodlice frequently live in the drainage holes of pots. When they are prevalent, it pays to stand each pot in a strong solution of insecticide for an hour or so. I have used for this a tin tray, holding several pots at once.

There are other chemicals for varying pests. Those that infest greenhouses, once they are introduced, are red spider mite, mealy bug and scale. They can all be eradicated by malathion. This is dangerous stuff if you are allergic to organo-phosphorous chemicals or the greenhouse is part of the residence. However, I know of one lady who has used it for many years in a small conservatory-like greenhouse with no apparent harmful personal results.

Malathion is one of the phosphatic chemicals that is most potent for several hours, but after two days completely disperses. It is not dangerous to outside life like DDT or even many of the so-called safe insecticides, such as nicotine. If used with the utmost care, it is, in my opinion, better than struggling with other pesticides that only do half the job. But it should be emphasised that other organo-phosphorous compounds of much stronger and more lasting powers must on no account be put in a greenhouse.

Beyond any doubt, the enemies of plant life most difficult to control are fungi of various sorts, some causing quick-spreading moulds. Most antidotes contain chemicals that stain the plants. If the warm greenhouse or stove is kept nicely moist, and the air is circulated, fungi rarely occur. If they do, it seldom pays to fight them. When a plant is badly attacked, or is in any way afflicted, it is best to destroy it. Not only does it become a thing that ceases to be beautiful, it is a menace to the others. And as healthy tropical plants grow quickly and can be propagated so easily, a vacancy can soon be filled by something nearer perfection.

These, broadly, are the principles of warm greenhouse gardening. In the next chapter we will go into a cultural programme in more detail regarding the propagation, care and attention needed throughout the year.

Especially to me the chief point is, not to become bogged down with a lot of extraneous details. The principles should be mastered, and after that, what you do becomes a matter of common sense, rather than the adoption of gimmicks. Once 'you know what you are doing', you can pursue your own individual ideas, rather than blindly follow a set of rules that you do not understand.

CHAPTER 2

Cultivation and Care Month by Month

HAVING grasped the broad principles of growing exotic plants in warm greenhouses, we can confidently give them the varied attentions required according to the season. Let us start at the beginning of the year.

January

This is a very quiet month, yet there will be flowers, if we grow subjects like eranthemums, and there will be several plants displaying their magnificent leaves in varying delightful hues.

Our main concern, though, throughout this month and the early part of the next, is keeping out frost, by maintaining as far as possible a pleasant, equable temperature. The normal one for the stove should be around 21°C (70°F) and 5 (10) degrees lower in the warm greenhouse. Temperatures should not exceed the normal by much and, as far as possible, should fall little below it, although this may be inevitable during a very frosty period. But we must remember that when cold, the plants suffer. On the other hand, if they are stimulated into premature growth by too much warmth, they will not develop into buxom, colourful specimens when the time comes.

Heating is controlled by keeping the fire burning so that the hot-water pipes emit the requisite warmth for the temperature required. It stands to reason that the more pipes there are, the easier it is to maintain that desirable equable state.

A good thermometer that registers minimum and maximum temperatures is imperative. It should be placed about a foot from the roof glass in the middle of the greenhouse, where it can be checked from the pathway. Its chief importance is to indicate the fluctuations of warmth and so help us to increase or decrease the heating.

When heat is thermostatically controlled it most often changes the flow of oil fuel to the furnace according to the water temperature at the source. In winter this has to be raised, especially in frosty times, for cold will penetrate through the glass quite suddenly if there is insufficient warmth to repel it.

Should the heating be electrical, this will be controlled in the greenhouse itself. When we set the thermostat, not only must we think of the inside temperature, but also make allowances for the penetration from outside.

During very frosty weather the sun will often shine sufficiently by day to warm up the greenhouse. However, we should never open any ventilators before the latter part of February. The air that penetrates between the glass laps, and through the door when it is opened, is enough.

Generally, the atmosphere will also be damp enough throughout the month, but should there be a feeling of dryness on a bright sunny morning, and the surfaces close to the warm pipes also are dry, then it will be good to sprinkle them with water taken from the inside tank, or to inject a little mistified moisture from the fine-nozzled syringe into the atmosphere. This must be done before noon.

As far as I know, there is no gadget that can tell you when to damp down. There are moisture meters, but they do not differentiate between the moisture content desirable at the varying temperatures. And in any case, directly you sprinkle water around, you change the atmosphere. If the air feels dry, give some moisture. If it feels clammy, be careful not to. If there is a feeling of buoyancy, it is inadvisable to sprinkle more than a little water around.

Watering. If you study the information regarding the differing needs of the various families dealt with in the main part of the book you will see that some need complete dryness during winter, others need to be kept moist and others need more water. Some are bulbs. Some are deciduous. Some are evergreen. The evergreens show when they want water, but the deciduous ones do not; if in doubt, a shrub that loses its leaves in winter will most likely need its roots to be kept watered.

Treatment for evergreens with the large, lovely leaves is to allow the compost in the pots to get almost dry and then to fill to the brim. If this is done, they will probably need no more than two such waterings during the month. They may indicate, by the flagging of the leaves, when they desperately need it. It is best to look at the compost almost every day, to make your own judgment.

When watering, the water should be poured into the top of the pot gently, rather than in a gush from a long-spouted can. One which has a short spout is much better. You can then control the flow.

There should be no shading, no syringing of the foliage, and no feeding with fertilisers during the whole of the month.

Pests are most unlikely. However, if the backs of the ornamental leaves assume a dirty grey look, you can suspect either red spider mite or thrips. You cannot see them. Obtain some liquid derris, dilute to instructions, add a dash of kitchen sink washing-up liquid and gently sponge the backs of the leaves. Or if too irksome, spray with a fine-nozzled syringe.

If any spring-cleaning of the house, such as rewashing the walls, or repainting the woodwork, is considered necessary, this is the best time of the year to do it. The majority of the plants are resting, and less moisture is needed.

February

By the middle of February the worst of the outdoor winter will have gone and the lengthening days will be more apparent. But, although we are away from the cold winds, or fog, or whatever comes in this unpredictable month, when indoors we must gauge its effect. If the cold persists we should be in no hurry to increase the temperature, and so incite plants to spring growth.

I go into the greenhouse in the morning, whether it be stove or warm, to check the thermometer, especially the position of the minimum indicator that shows how cold it has been during the night, and indicates how the place has warmed up since. I then always adjust it to watch the progress of the temperature as the day passes. When it is warming up nicely to 21°C (70°F) in the stove, or a few degrees lower in the warm house, the top ventilator is opened for the first time during the year. Just a crack of air is allowed, and that is shut off again soon after noon. By the end of the month, depending upon the outdoors weather, and particularly the amount of sunshine, it will be opened for three or four hours, the aperture never being wide at any time.

The pots are all examined, and water given to those that have become dry. And if the air seems dry also, some water is sprinkled on the pathways, and in the region of the warm pipes. There should never be enough to create a clammy atmosphere, but rather a close, moist one, pleasantly warm. No water should be sprinkled around after the early afternoon. We are still conserving the artificial heat from the apparatus as much as we possibly can. If a fan is one of the fittings, it is turned on and left running. It circulates the air and deters moulds from forming on the structure or the plants.

Seed Sowing. The wonderful flora of the tropical countries produces the most amazing contrasts in seed, which is the most natural means of propagation. There are coconuts wrapped in coir that are floated hither and thither during floods, while at the other extreme is seed as fine as dust that is blown about by the winds. We do not raise palm trees from coconuts, because they are not exactly things of beauty for our greenhouses, but we do raise some of the finer-seeded plants that have dust-like means of reproduction, chief of which are begonias and gloxinias. The best time to do this is late February.

Pans, either clay or plastic, are filled with one of the no-soil composts. Alternatively, we can mix silver sand, rather than horticultural grit, with fine sedge peat, or we can buy John Innes seed compost for the purpose. I have found a good brand of no-soil seed compost, consisting mainly of Irish moss peat, most excellent for this. There is no need for special drainage, but if John Innes is used, some rough peat or dried oak leaves should be placed over the drainage holes in the bottom of the pans before it is put in. Whatever compost is used should be tamped flat with a board or a brick, and then be well soaked with water and placed in the warmest part of the greenhouse to drain. A few hours later the seed is sprinkled over the surface as thinly as possible. It is not covered by more compost or anything similar but, instead, a pane of glass is placed over the pan and over that a sheet of newspaper, or cloth. Each morning, when doing the usual round of watering, the top covering is lifted off, the glass is turned over so that the dry side is now put underneath, and the cover returned. In due course, there will be signs of germination and then the top covering is removed completely. After a few days the glass is also taken away.

A good position for the seedpans during the germination period is the floor of the propagator. Alternatively they could be placed near the heated water pipes or, if an electric stove is used for heating, close to that. But drying out is likely to occur in too dry an atmosphere. Should the seed compost ever get very dry during this period, it is never watered upon the surface, only by steeping the seedpans in other shallow pans of tepid water, so that the compost becomes thoroughly moistened by seepage through the drainage holes. The water surface should always be below the seedpan compost surface, or it will rise up and float the seeds away (see Fig. 3).

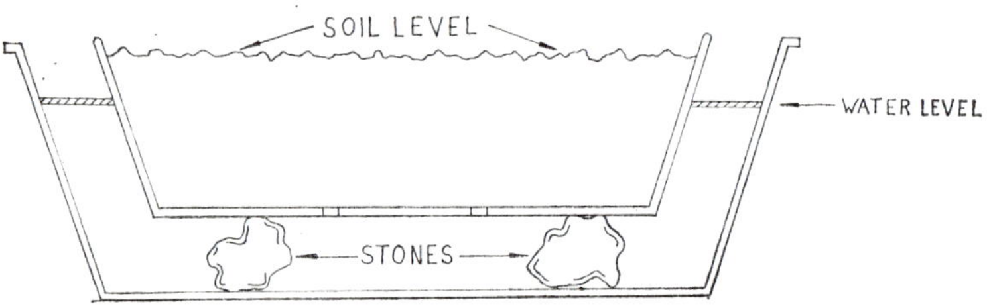

Fig. 3. Steeping seedpans.

The seedlings have to be transplanted while very tiny. They are generally overcrowded and, if left to grow, would soon be spoiled by intensive competition. Transplanting will probably be early March, depending upon the time of sowing and quickness of germination. They can either be potted off singly into tiny pots or 'pricked-off' into other seedpans, which is my favoured procedure. For this either use John Innes seed compost, a no-soil compost, or your own mixture of soil, sand and peat. I prefer mixing Irish peat with the standard John Innes potting compost. Seedlings grow more freely in no-soil composts at first, but soon exhaust the space limits and start to deteriorate. On the other hand, they are slow to start in John Innes but continue to develop sturdily. By mixing in peat constant growth throughout is obtained.

When the compost is put in the 'pricking off' pans, it should be thoroughly watered, and then allowed to dry sufficiently to enable one to poke it about with a stick without any adhering. A peg, somewhat thicker than a pencil, is used to make the holes and also to loosen the seedlings in the original pan. The hole is made dibber-wise, and the seedling then held suspended with its roots in the hole, gently covering them by pressing soil over them with the end of the wooden peg. Instead of using the fingers to handle the seedlings, a pair of tweezers makes the job easier. Spaced an inch apart, they have plenty of room to grow. Do not properly water for a few days. Put a shade over them if the sun can shine upon them. But if they droop, they should be moistened by gently spraying with a mistifying syringe. The generally genial, moist atmosphere of the greenhouse will help them to recover completely from the operation within a day or two. Eventually they are potted in the normal way.

Tubers and Corms. If you have old begonia tubers that have remained sound and healthy during winter, you now place them in a shallow box of sedge peat or, if you have it, well-rotted leafmould. It should be damp, and barely cover their tops. Placed in the stove, or the warm greenhouse, in a semi-shady place, they will soon form roots, and can then be potted singly into sizes that will only just accommodate them. Having got them to this stage, they are best removed from the warm greenhouse into a cooler one that is out of the realm of this book.

The beautiful gloxinias can also be started from tubers but, instead of treating them like begonias, they are usually planted at the outset in small pots. This can be done throughout February or, if you like them late, during March. If a stove house is available, they can generally be started early. If they must only be grown in the warm greenhouse, they are best left until later. They and the begonias, when potted up, will grow in John Innes-cum-peat compost. This will be when the plants have formed good roots. Later, they are transferred to larger pots, the 5-in. size being ample at this time. The gloxinias always need warmth.

Some achimenes tubers can also be started by planting them in pots, or hanging baskets. They grow well in the same compost. The treatment slightly differs from either of the above, it being usual to plant them more like bulbs, by putting several in a receptacle, spacing them just over an inch apart. They need to be about 2 in. deep. After the compost has been soaked, and then partly dried, and the tubers planted and covered with more compost no more water should be given until growth is visible. After that give more; sparingly at first, but more freely as they grow. They are not potted-on like the begonias and gloxinias. They can be increased by dividing the tubers at this time, or by sowing seed.

These are the main three sorts of plants that need this type of treatment. Among others are gesnerias and caladiums. More information about them is given in the lists.

Pruning. A study of the cultural hints heading the particulars of the various families, and their genera and species, will reveal that some plants need to be pruned back annually at this time of year. Others need no such treatment. In some instances it is arbitrary, depending upon one's needs.

It may be wise to shorten the branches if they are required to sprout new shoots that can be taken off for cuttings. For some it may be necessary to do this pruning during February, in order to get new shoots for propagation purposes.

Pests. With the awakening of the plants, the lengthening of the daylight hours, and the increase of heat and moisture, there may be lurking insect pests about, especially ants, woodlice and just a chance of aphis. With more moisture, red spider mite is less likely, for it cannot thrive in water. The chief thing is to be observant.

If there are resident mealy-bug or scale insects present, they will now be moving and breeding, and must be eliminated in the most efficient way possible. Their presence is unmistakable. The bug not only clothes itself with a white woolly robe, but also exudes a sticky liquid. The female lays eggs in the wool, which hatch into youngsters that crawl about the stems and get into the joints of the leaves.

Scale looks somewhat like a smallish edition of an undernourished, bleached ladybird, that adheres to the stems. It and mealy-bug can be sprayed with a malathion wash, taking care not to inhale the fumes or stay in the vicinity until after several hours have passed. An aerosol is the best means of application, if it can be obtained. If not happy about this you must seek other methods of extermination.

March

The daylight hours are quickly lengthening, and the temperature in the stove and warm house soars when sunshine prevails. It is kept down to between 21°C (70°F) and 27°C (80°F) by opening the top ventilators just before that amount of warmth is registering, closing them in the early afternoon, so as to keep in as much warmth as possible during the nights, which if cold, means giving a full amount of artificial heat, by whatever system the greenhouses are kept warm.

Plants need watering more often. Those with leathery leaves like to be syringed. And damping down by sprinkling water around, is liked, especially close to the source of artificial warmth.

Those plants with hairy leaves, such as saintpaulias and the other gesneriads, should not be syringed. They love the humidity, but not solid water. An automatic mistifier placed somewhere near them will be beneficial.

It may be advantageous to start an automatic moisture dispenser at this time. But this should be done somewhat tardily. It can be most valuable throughout summer, saving a lot of irksome attention, but in spring and autumn may possibly cause clamminess, instead of giving buoyant humidity. It should never be used during winter.

Propagation. As stated in the first chapter, there are two methods of managing a propagator. One is enclosed, while the other has an open top, the cuttings being kept turgid during the rooting process by frequent intermittent injections of mist overhead. If we adopt the mist propagation unit, rather than the enclosed one, this will spread into the general greenhouse atmosphere and reduce the necessity to damp down frequently, yet not cause too much dampness during this crucial month. It is all a matter of judgment and experience.

The most usual way of increasing tropical plants is by cuttings taken off old plants at various times of year, according to their habits. Some are so obliging that it can be done at any time, although it is inadvisable to do it during winter.

Actually, during early spring and late autumn, the enclosed propagator is better than an open one. However, from the middle of March onwards to early autumn, what with the humid heat of the greenhouse and the special dampings given by the mistifier, most subjects root more easily (see Fig. 4).

Normally, the best cuttings are new shoots about 3 in. long, that are neither too soft and lush nor weak, but look good. They are pulled off the plant with a small piece of the old stem attached, or, as we say, a 'heel'. Just the bottom end of the stringy bark is snipped off with an old razor blade or very sharp knife and also the lower leaves, if any, so that the cuttings can be inserted to half their lengths without burying any appendages.

Some gardeners place them in a bed of sand and peat, or

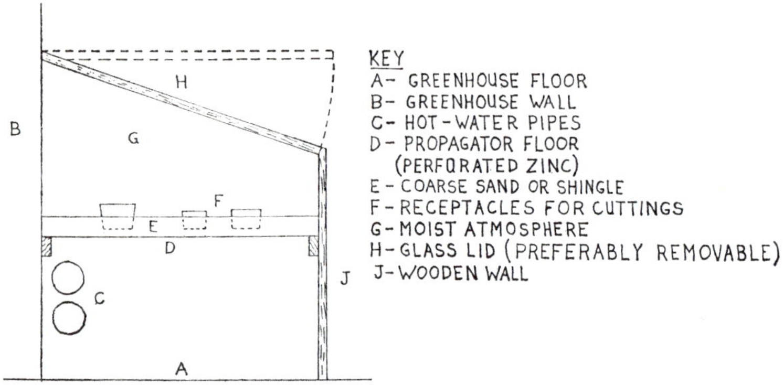

Fig. 4. Home-made propagator for pipe-heated greenhouse.

concoct their own supposed magical rooting mediums, and either prefer to place them in pans or pots, singly or in multiples. Sometimes I have used silver sand and sedge peat in equal proportions. At other times I have used vermiculite and moss peat. I have always preferred to use receptacles rather than put them in the warm bed. They are stood on the bed or slightly dropped in the sand over the cables or pipes, so that they get the desirable bottom heat.

The best receptacles are 5-in. clay pots, filled with the rooting medium, and then saturated with tepid water, which is allowed to dry out until the compost is easy to handle. Four cuttings can be placed in each pot, spaced evenly to form a square. They are inserted close up to the inside surface of the pot itself so that the bottom leaves overlap the rim. This may just be fanciful, but many of us believe that they root better like that, than if they are inserted clear of the sides. Once inserted they are placed in the propagator which, of course, keeps them moist above and below.

After being in the warm moisture for three or four weeks we can check to see if they have rooted. By lifting a pot and carefully holding it upside down, and gently tapping it on the edge of the staging, with one hand under the top, the bottom can be lifted up, revealing the ball of soil with the visible cuttings. If roots have formed, they should be put singly into small

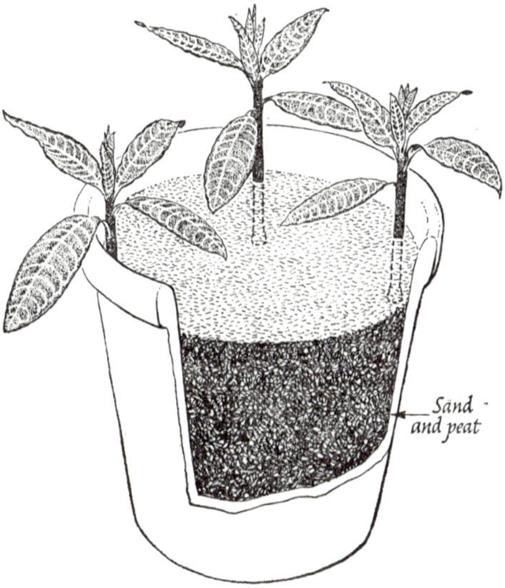

Fig. 5. Cuttings trimmed and inserted against inside of 5-inch pot.

pots, using one of the branded no-soil composts, and placing them where they get warmth and a fair amount of moisture.

Some plant species may be tardy in rooting. In that case it may be advantageous to use one of the hormone rooting powders. But root inducers are not always successful. If cuttings of plants root easily without them, there is no need to bother. If there is difficulty, they can be tried. But most stove and warm greenhouse plants are so obliging that one should find out first what is being done wrong if they are generally reluctant, rather than resort immediately to chemical aids.

Sowings. If there are any seeds of plants that need exceptionally warm beginnings, they can be started now. Most often, we are too busy during this month to involve ourselves in any additional responsibilities. But it is possible to sow them now. If the seed is very fine it should be treated in the way described for gloxinias. If it is larger, the general guiding rule is to cover it with very fine soil, or very fine peat, or peat mixed with silver sand, so that the seed is covered with twice the depth of its own diameter.

Very hard seeds can be softened by soaking overnight in tepid water; for some the skin is filed upon the side farthest from the 'belly button'.

Potting. We also have potting to do during March. There are two distinct forms of this. In the one we 'pot-on'. In the other we 'repot'.

First of all, the potting-on simply must be done as it arises. It is done when quick-growing subjects like begonias and gloxinias, raised from seed or started from corms and then potted in small sizes, need more room. The same also applies to rooted cuttings. It is absolutely essential that they are only put into the smallest pots possible at the beginning because, if put directly into the larger ones, their roots only half use the compost, the rest becoming sour and so unsuitable for them to spread into.

The other form of potting, or repotting, is changing the compost for old, established plants. This need not happen every year. In fact, it is inadvisable to do it other than when necessary. The plants themselves tell you when this is—they look listless, and seem to be feeling sorry for themselves. Sometimes they need to have the stems or branches shortened at the time they need to be repotted, which should be some time in spring.

Heeding their preferences for compost mixtures by checking encyclopaedias, if they are not given in my lists, tip the plants out of the old pots, carefully prodding out as much of the old compost as possible. Then cut off any roots that are obviously dead, taking care to retain every piece that is living, and repot carefully, holding the plant at the right depth with one hand and scooping compost in with the other, pressing it among the roots carefully but firmly with the fingers. Having got the plant to the stage when it can stand upright, both hands can then be used to put in more compost, again pressing it down and around with the fingers.

Old gardeners are scornful about using thumbs. They should always be kept upon the outsides of the pots.

Before potting any plant, give it a good watering two days beforehand. Then it is generally in the right condition to handle. To tip out, put one hand across the top of the compost, with the stem of the plant between the fingers, turn it upside down and rap the rim on the edge of the bench. The plant will generally come out into the palm of the hand with the ball intact.

The compost used for repotting can be plied more easily if in the right state. When too dry, give a thorough watering and then allow it to get to the drier damp state. If too wet, leave it for a while. The check for readiness is to take a handful off the heap and squeeze it tight: it should stay intact, but when dropped back on the heap, should disintegrate.

Old plants must never be put into too large sized pots, any more than young ones should. Often as a result of removing the compost and pruning off dead roots, it is sometimes advisable to use a smaller size than the original. But once they fill it, the original size is again necessary, or a larger one.

After potting it is best not to give water immediately. In fact, it is better not to do so for three or four days, but in the meantime put the plant in a warm, humid, shaded corner so that the foliage does not flag unduly. It is in reality more or less a half-rooted, aged cutting, and must produce new roots in compost that is not sodden.

April

Increasing daylight; increasing heat; increasing interior moisture; increasing ventilation; increasing growth of plants; increasing colour and redolence; increasing pest troubles. These are the chief happenings during April.

It can be a treacherous time. The temperature can rise outdoors by day, and the bright sunshine make the indoor temperature soar. By night, there may be very sharp frosts that generally come with the arrival of the orchard apple blossoms. The chief thing we must not do is to allow ourselves to be gulled that summer is here.

Only the top ventilators should be opened during the day.

We water the pots that need it, as early as possible in the morning, syringe the growing, leathery-leaved plants, and if the sun is rising fast, give a crack of ventilation at the top. But if it is a dull morning we defer this. We even go steady with the syringe. We listen to the weather forecast overnight, and make up a brisk or a cool fire accordingly. Should we have thermostat heating we turn it on a little if there is a threat of frost, not forgetting to turn it back again in the morning, for almost always sunshine follows, which runs up the temperature again very quickly.

About midday, we sprinkle water around (damp down) to increase humidity and prevent a dry atmosphere from forming. In late afternoon, we syringe those plants that can take water once more. We close the ventilators about two hours before sunset, or earlier if it is cold outside.

Shading. In some years it is necessary to provide shading during the last half of April. But not every year. When it is bright and sunny is the time to think about it. Most tropical plants love the light, but most are protected from the blazing sun by other taller and stronger growing vegetation. Undoubtedly the best way to provide shade is by easily controlled blinds that can be pulled up and down as deemed advisable. Otherwise, a shading paste should be used.

Labels. All plants need labels, for no one's memory is foolproof. And there is nothing more tantalising than not being able to tell another person the name of the plant that is being admired. There are innumerable types, made in all sorts of materials, the commonest just now being of white plastic. But I do not think them ideal for warm greenhouses. Unless the names are written properly, they cannot be read easily. And most of them have shiny surfaces difficult to write upon. There are varying sorts of fluids to write with, and also special pencils.

The labels that I like best are still the old-fashioned wooden ones, painted white on the side that has to be written upon, using a genuine Horto Garden Pencil, that does not wash off. It is a labour of love to go round the pots occasionally to give new labels where needed. Alternatively, there are metal labels with the names embossed on them by the manufacturers, or

▲ *Crossandra infundibuliformis*. Dwarf evergreen shrub. Flowers are orange-red, arriving in spring.

◀ *Dracaena* 'Victoriae'. Small evergreen tree, chiefly grown for attractive foliage.

▲ *Euphorbia pulcherrima*. Well known 'poinsettia' with usually scarlet but sometimes pink or white bracts.

▼ Gardenia. Evergreen shrub bearing intensely fragrant flowers intermittently.

you can obtain your own machine and do it yourself. Actually, the practical approach is to have wooden labels for seedlings and young plants, and then, when assured they will become permanent occupants, honour them with the embossed ones.

Anyway, April is a good month in which to check through them all.

Feeding. By this time, some of the plants potted up earlier will have used up much of the goodness in the compost, especially it it is of the no-soil type, or if plain peat has been added to John Innes potting compost to lighten it, and will require extra feeding. Also, old plants that were not repotted, if they are healthy, can be given a little fertiliser.

The most important point here is to be able to distinguish a plant that is unhealthy from one that is merely being starved. An ill plant cannot take fertilisers. A healthy one can.

Another thing we have to do, to be first-class plantsmen, is to know when not to give thriving plants very much additional fertiliser because, if we do, it makes for soft stems and leaves instead of firm, colourful ones. Therefore, it is bad to feed them all willy-nilly. We try to understand our plants and let them tell us when they can take it.

As before stated, there are several good branded liquid fertilisers available, often based upon maritime seaweed extractions. If the instructions upon the containers are heeded, it will be found that a little goes a long way.

Propagation. Plants can be increased by the various methods of seed, cuttings, or leaves as described on foregoing pages. But at this time of year, it is possible to become overcrowded and it may lead to having to make decisions to discard this or that to make room for this or that. Jumble Sales and other 'do-good' forms of aiding others always appreciate your unwanted plants. There will surely be enthusiasts willing to contribute to such funds in order to own lovely plants.

Pruning and Training. If there are climbing plants, many need to have the branches shortened or thinned out, or to be trained into the ways in which you want them to go. When doing this, it is important to know where the flowers will appear; so many plants have them at the ends of the current growths. In that case, once the framework has developed, only excessive branches should be cut clean away down to a fork. For tying in, my preference is green raffia. Too thin string strangles growths.

Pests. While pests could be most troublesome, they need not be, provided the stove and warm greenhouses are attended to properly and precautions taken, such as ensuring that no pests are introduced by putting new plants in quarantine first.

However, woodlice and ants may appear. Ants will actually carry greenfly into the greenhouse to start colonising if they have the opportunity. One of the proprietary antidotes is the best way of coping with them. For a general insecticide, I favour the careful use of malathion. It can be used in the form of dust or liquid, or as a fumigant. Probably the most suitable and easiest form of application is by an aerosol. It is the push-button method of destroying local outbreaks of mealy bug, scale, aphids, thrips, red spider and whitefly. Probably Sevin (carbaryl) dust is the best antidote to kill off woodlice.

May

Practically all that has been written concerning April can be repeated for May. Watering becomes more frequent. But one can take consolation from the fact that when the atmosphere is filled with moisture, there is less evaporation from the pots. They need looking through daily. One likely fault is to soak the lot, in which case, a plant that is not using the water to that extent will suffer from lack of drainage and generally get worse, instead of benefiting from watering. If a valuable specimen is being so affected, it should be repotted, so that it gets more drainage and new compost. Should it be top-heavy with foliage, do not hesitate to shorten it. Then, when it sends up new growths, take some off for cuttings in the usual manner.

One must still keep the heat going, if ever so slowly, for nights are often cold during most of May. At this time, if there is an electric system, a heater turned on late at night and switched off in the early morning is all that is necessary. By the end of the month, even this should not be needed.

Damping down now becomes an art, as indeed the whole management of a stove or greenhouse is, all the time. Open the ventilators when the temperature soars. Close them again when it threatens to drop. But do not expose the plants to too much air that will dry them out. Stoves need less ventilation and fewer ventilators than warm greenhouses, and warm greenhouses less than cool ones (the notes in the chapter about orchids also deal with this aspect).

June

This is a glorious month in the stove. It can also be most showy in the warm greenhouse, especially if some of the stove occupants have been moved into it, if one has both. Ventilators need to be opened in the early morning, and the blinds pulled down before the sun becomes scorching hot. It does not matter if the temperature soars to 30°C (86°F) at its zenith, for by damping down it can soon be reduced. If there is an advantage in shading with a paste it is during June, because the sun rises so early, often long before we awake.

There is no need to apply any artificial heat. Should there be a few cold nights, they can be circumvented by closing the ventilators fairly early to trap the warmth. The nights are short and before the cold has seriously penetrated, the sun is rising again.

July

As for June, also notes for May apply. Watering, spraying, damping down, looking for pests, tying-in and pruning exuberant plants, and enjoying the show of leaves, flowers and fragrance all make this a most wonderful month.

There are some plants, such as poinsettias, that prefer somewhat cooler conditions. They need to be put where they can get them for the next few weeks, in order to ripen their shoots that will not bear the bracts so grandly if they are kept soft.

August

The ripening or toughening process of several plants is now necessary if we are to keep them up to standard. More air is given. This means a slightly drier atmosphere inside. If we get a rainy spell, which we sometimes do during this month, the blinds should be kept rolled up, and if there is paste shading it should be washed off. It is also a great advantage to have an electric fan inside the house to keep the air circulated. It is better than opening lower sashes of warm greenhouses.

Holidays. There may be a problem this month of keeping the plants alive or in good condition while the family is on holiday. It is tempting to obtain one of those watering systems that feed the pots by capillary attraction. There are several systems, but I am not qualified to comment upon their respective merits. They often go wrong. Also, for me, they detract from the general appearance of the ornamental house filled with such beautiful subjects. I want my friends to concentrate their full admiration upon them, and not be diverted by any artificial contrivance, no matter how clever it is.

The fact that I favour the mist dispenser in the propagating frame, and appreciate electric fans and thermostatically controlled heating shows, I hope, that I am not totally averse to being helped by modern aids. These do not mar the general picture of the house and its occupants. But wet, slushy benches and pipes and spouts that snake in and around the pots are too diverting. Notwithstanding, I do give further information about them in Part Three.

If one is away for a week or ten days and it is impossible to call in a reliable help, the plants can be kept in good condition by lifting all pots from the staging and placing them closely together on the floor. Some can go under the staging. If there is not enough room, some can be carried to another room or outbuilding. All should be plunged into trays filled with peat. Now that plastic ware is exceedingly cheap and light, enough large shallow receptacles can be obtained without much outlay and retained for this purpose if there is no other source for them in the interim. Put into each enough Irish moss peat to half fill them, and then pour in water so that it is thoroughly wetted. If there is any surplus water, drain off by holding the tray on its side for a few minutes. Then place the pots in the sloppy mixture, making sure they themselves also have been watered. Put as many as possible in each tray, and stand on the pathway, or anywhere convenient. If there is a mist dispenser allow it to keep working throughout. Give just a little air by slightly opening the top ventilators. Put the paste shading on the glass roof, but do not drop the blinds, unless they are of a sort that will allow in as much light as the shading would do. In that case use them instead. The chances are that these plants will be in good condition when you return, but they must be put in their rightful positions again as soon as possible. Any damaged or decaying leaves must be removed.

September

The nights are beginning to become noticeably longer and there is not the urgency that there was of opening the ventilators in the morning. Nevertheless, they should be opened during nice weather. Watering will be less urgent, and there will be plants coming to a state of rest, when it has to be withheld altogether. The mist dispenser in the propagator should be turned off for the winter. If there are any cuttings to be rooted they are best put under glass.

We aim to get the main collection of plants comparatively toughened so that they can stand the oncoming winter all the better. The house is kept moist, but not as much as in the summer. It may even be good to restart the fires temporarily during inclement periods. The top ventilators should be kept open by day when this is necessary. We stop giving fertiliser to any plants. The temperature by day is gradually getting lower. The blinds or the paste shading should be removed. It is unlikely that they will be needed again until early spring. Any prevailing sunshine will be welcome unless it is excessively warm. It seldom is.

October

During the first half of this month we give quite a lot of top ventilation and also provide artificial warmth. Watering becomes less urgent. Syringing is carried out only where deemed necessary. We are trying to get the plants mature and fit to withstand the oncoming winter. There must be enough moisture in the atmosphere to stop red spider mite encroaching. We are striving to balance summer with winter.

By the end of the month the caladiums, tuberous begonias and gloxinias will be going to rest. When they have completely lost their foliage they should be packed away in a dry place where they get warmth, not wet. I prefer to keep them in their pots, while other growers often take them out and store in bags, or in sand.

If there is an electric fan it should be kept going. It circulates the air and drives away any moulds or mildews that will arrive if there is clamminess.

November–December

The opening notes about January are also applicable to these two dark months. Frosts are less severe than they are during January, but we must guard against undue dampness. The air should be moist, but never laden with water. The fan and the heaters will counter that problem. We can always sprinkle a little water around if it feels too dry.

These month-by-month hints are general. By studying the particulars about the various sorts of plants in the succeeding pages, you will find many have specific needs, or differences, from the foregoing. A compromise may sometimes be necessary. Withall, the good grower is never precise about anything. But by a little thought, and a desire to understand, one can soon acquire 'a feel' for the different subjects that matters more than any rule-of-thumb.

Part Two

CHAPTER 3

The Plants

SOME of the generic names of the plants that live happily in the stove or warm greenhouse managed in the normal way are quite familiar. However, there are many that are not, and to list these in mere alphabetical sequence of genera, as is usual in encyclopaedias and dictionaries, can only convey bewilderment. However, we must introduce them somehow, and I do by, admittedly, using an alphabetical list, but of botanical families.

I have always found it difficult to remember names when I want to recall them most, but by adopting an emotive method, rather than one of formulative correctness, I manage to trace almost any title I want when it becomes necessary to do so.

Of course, we have labels. They get lost or changed. Or we buy plants that are not true to name. However, by practical application of a basic knowledge, it is possible to know when someone is right or wrong. And if we pursue the subject, it becomes fascinating.

The first thing to do is to be able to know to which family any plant belongs. Not only does it help to decide the treatment it will need, but it also informs us about its future behaviour. There are exceptions, and we do make mistakes. However, they will be far fewer if we adopt this family approach, rather than try to remember the plants one by one, regardless.

I regard the aroids as being the most important of them all because they include such a vast number of variable and exciting species ranging from arum lilies to monsteras, including the gorgeous caladiums whose leaves are truly exotically tropical, to the utmost extent. Following the aroids are the bromeliads, or pineapples, that make magnificent plants, and so on, until finally, the lists are completed with miscellaneous individuals that have few or no relatives in a collection. In some instances, I have never seen living specimens, but have relied upon illustrations and descriptions taken from the works of other gardeners or writers. But just because I have had no personal experiences, I must not refrain from including them in my lists. It would be unfair to the plants.

Ferns and palms are welcome additions to collections of ordinary plants, and I have also considered a few orchids that can be grown along with any of the main subjects.

But there are many desirable species and varieties omitted because it is practically impossible to mention them all. The chief aim is to make the reader aware of facts so that he or she can pursue any family or group of plants if the interest is awakened to that extent.

Where I refer to genera which I consider to be of lesser importance, these are separated by asterisks from the main subjects in their respective families.

CHAPTER 4

The Acanthads (Acanthaceae)

ACANTHUSES of the herbaceous border, sometimes named bear's breeches, carry the family name of many genera in which there are over 2,000 species, a great number of which need a stove or warm greenhouse in order to live in this country.

Probably the most beautiful and useful of all are the aphelandras, which have very brilliant variegated foliage in exciting, shiny art shades, while the bright orange or yellow flowers, which arrive during late summer and autumn, give the most impressive splash of colour at a very convenient time of year.

The popular shrimp plant, *Beloperone guttata*, is an acanthad that is easier to grow than most, it being happy in a cool greenhouse or a living-room, although it will live in a warm greenhouse. But it is best in a cooler place, for if kept too warm the colour of the bracts is not good.

Still another well-known member of the family is *Thunbergia alata*, often called black-eyed Susan, growing well as a twining annual in a cool or warm greenhouse.

Aphelandra

However, as the evergreen aphelandras set the general pattern of the main stove or warm greenhouse species, their cultivation must be given first consideration. Apart from timing, the majority of other tropical, or temperate, acanthads will thrive under similar treatment.

They grow comparatively fast, and start flowering very freely quite early in life. Consequently they can be had as small plants by raising afresh every year or, if kept, they will develop into sizeable specimens as the seasons pass.

The normal beginning of an aphelandra is as a cutting removed from an older plant during April or May. After the old plants have had a late winter rest, and have started again, some new shoots are pulled off the stems when about 3 in. long, keeping the heel parts. They are lightly trimmed, and then either put into small pots individually, or with companions in a pan. The compost mixture should be half Irish moss peat and half sharp river or silver sand. After insertion they are given a nice soaking of tepid water from a fine-rosed can and then placed in the propagator.

If the temperature is maintained at a steady 21°C (70°F), good roots will soon form. They will be quickened if the propagator has under-floor heating (bottom heat). In the normal way they will be taken from the old plant in late April and will be sufficiently well rooted to be taken out of the propagator by early June.

They should be kept in a moist stove house, and syringed as frequently as they seem to need it, depending upon the weather. If the sun is bright, they must be shaded, preferably not all the time, for although direct sun can damage, they simply must have plenty of light.

They soon fill the small pots with roots and then need transplanting to slightly larger ones, the 5 in. size being suitable. For this, the John Innes potting mixture is good, but if they are to be grown for only one or two seasons some sedge, or moss, peat can be added to accelerate development. But exclusive, patent no-soil composts are not advocated. They are inclined to make for too fast growing at first, to the detriment of the colourings of the leaves and brilliance of flowers.

If kept moist and warm, and lightly shaded when the sun is shining, they grow quickly and some may benefit if repotted into slightly larger sizes.

One important point is to pinch out the growing tip of the main shoot to induce bushiness, and it may be advisable to treat other straggly shoots the same way. It may also be helpful, in some instances, to insert canes into the pots to train the shoots by tying them outwards.

When the flower heads appear in late summer, the roots should be given a little liquid fertiliser. One of the proprietary brands, like Liquinure or Maxicrop, will be most suitable. If there is a warm house, they can be transferred to that from the stove until March. But it must be warm, as should a conservatory, or a room with a nice light window, where direct sun is avoided. Wherever they are at this stage, do not give more water than is absolutely necessary, which means allowing them to get rather dry before soaking the roots, and then refraining until they need water again.

In February, prune hard back all the side-shoots and shorten the main stems considerably. New shoots will soon appear and when 3 in. long are suitable for cuttings, as already mentioned.

However, if you wish to retain the old plants so that they grow into specimens, after pruning, they are repotted into John Innes potting mixture to which peat is added. When tipped out of the old pots the soil forming the root ball can easily be rubbed out without injuring the roots. If there is a straggly root it can be cut off, but it is inadvisable to tamper with them very much. If they will go back into another pot the same size as the one they came out of, it is best to use it, or even a smaller one. Later, when fresh roots fill the pot, a plant can be put into a slightly larger one. Sometimes a little fertiliser at regular intervals in late summer will carry it through until flowering ceases.

APHELANDRA AURANTIACA ROEZLII. Makes a beautiful plant of smallish size with dark green and silvery leaves. Flowers are comparatively large and bright orange-scarlet.

A. CHAMISSONIANA. Has leaves with broad bands of white dots at the edges. Flowers and bracts are yellow.

A. 'FASCINATOR'. A beautiful species with variegated leaves in shades of silver and purple. Flower spikes are large and of a vermilion colour.

A. SQUARROSA. Has leaves about 10 in. long that are either green or green and cream. Flowers are yellow with scarlet bracts.

A. S. LEOPOLDII. A good grower with broad, shiny foliage, that has strong white veins. Flowers are yellow with conspicuous red bracts.

A. S. LOUISAE. Leaves are narrower and with more pronounced white veins. Flowers are saffron.

A. TETRAGONA. A strong, handsome grower that bears crested spikes of bright orange-scarlet flowers.

There are several other species, more or less rare, most of them autumn and winter flowering, needing identical treatment.

Crossandra

There are several species of this genus which mostly originate in tropical Africa, but the best one, next dealt with, is from the East Indies. They are closely related to aphelandras and need practically the same treatment. But the time for taking cuttings depends on the season.

CROSSANDRA INFUNDIBULIFORMIS (syn. *C. undulifolia*). This species has fairly packed leaves that are oval and wavy-edged, and flowers are produced around a spike, looking somewhat like lopsided primulas. They are bright orange, and continue for many weeks. The height is about 2 ft.

Eranthemum

At one time these comprised a largish section of stove acanthads, but have since been given other names that include *Pseuderanthemum*, *Odontonema* and *Xantheranthemum*. On the other hand, some originally under the name *Daedalacanthus* are now named eranthemums. Whatever their names, the chief thing is that practically all are very easy subjects in the stove, and are noted for freedom of flowering during late autumn, winter and spring, according to their kind. Cuttings root easily during spring and summer, and the best potting mixture is one of peat, leafmould and sand, or one of the no-soil proprietary brands. They love light, and should be kept near the glass, as the weakness is to grow leggy. Cuttings should be potted, and then re-potted as often as needed. They are rapid growers and always hungry, and therefore doses of liquid fertiliser are helpful. They need a little air to make them tough, but on no account should they be subjected to cold draughts. After blooming, the tops may be cut back, and the main plant kept just sufficiently moist during winter to live. In spring either repot, as recommended for aphelandras or, after taking cuttings, discard the old plant. Young plants are best.

ERANTHEMUM ANDERSONII. This is an Indian species (now named *Pseuderanthemum acuminatissimum*) that has white flowers, with lower petals dotted crimson, during November.

E. ASPERSUM (now *Pseuderanthemum aspersum*). Is a fine growing species with oval leaves and white flowers that are spotted crimson-purple, especially on lower lobe.

E. PULCHELLUM. This is the best known species and a rather strong grower. It produces flowers upon almost every part of the plant, that are dark blue on short spikes, and make a brave winter display. Should only be shaded during hot sunshine. It pays to pinch back straggly shoots, giving the plants more substance.

There are several other good species now placed in *Pseuderanthemum*, among them being: *P. cinnabarinum*; *P. malaccense*; *P. tuberculatum*.

Fittonia

These are dwarf evergreens considered to be difficult to grow in households, but quite easy in the stove, where they embellish the look of the other occupants, and grow in shallow bowls or pots, or even in compost placed on the staging. The general appearance is of dense foliage in the form of an attractive hummock. The flowers are insignificant and are best picked off in the early bud stage.

They like humidity, and prefer some shade. Little pieces taken off the best shoots and dibbled into shallow pans, or even on wet mounds of earth on the staging, will quickly root. They are useful for hanging baskets if these are needed in the stove house. The chief needs of fittonias are warmth and humidity.

FITTONIA ARGYRONEURA. This is a compact dwarf species, that spreads and spills over pots most gracefully. The leaves are delightfully marked, the basic colour being pale green, while the whole is profusely netted with silvery-white veins.

F. VERSCHAFFELTII. Has pleasing green heart-shaped leaves that are gorgeously netted with red veining. This is a trailer.

F. V. PEARCEI. Has somewhat paler foliage than the species, having more pink in the veining.

Ruellia

These comprise another group of shrubby acanthads. They are for the most part handsome, and respond to the treatment for aphelandras. They differ in their times of resting, according to season of flowering. They are propagated from cuttings, and some bear flowers very freely.

RUELLIA FORMOSA. Grows up to 2 ft. and bears many scarlet flowers during summer, needing an autumn rest.

R. HERBSTII. Grows up to $2\frac{1}{2}$ ft. and has long narrow leaves. Pink flowers are borne in autumn. Needs a winter rest, but should never be allowed to completely dry out.

R. MACRANTHA. The Christmas Pride is very similar to the last-named, but has flowers during winter and spring and therefore needs a summer rest, but should not be allowed to dry out completely. Flowers rather remindful of salpiglossis.

R. PORTELLAE. Has attractive, smallish, patterned leaves of a bluish hue, and rosy flowers in winter.

Strobilanthes

This genus is somewhat mixed and will probably have some name changes. It needs stove conditions, and treatment as for aphelandras.

STROBILANTHES DYERIANUS. The cone head has oval leaves painted rosy purple, slightly resembling coleus. The funnel-shaped flowers are a pale blue shade.

S. MICHOLITZII. Grows about 4 ft. tall. Has long narrow leaves, with white flowers clothed with green-tipped bracts.

* * *

THUNBERGIA ALATA. Black-eyed Susan is a perennial most often grown as an annual, it being possible to have flowers within three months of sowing seed. Colour is usually mustard yellow, with black centres, hence the common name. There are several varieties and cultivars much more variable, including a white one. The habit is to climb somewhat like a convolvulus, on a framework of canes or sticks. If sown in spring flowers are obtained during late summer. If sown in autumn they appear in spring. Use John Innes seed compost. Sow about six or seven seeds in a 5-in. pot. When large enough, pull out all but the three strongest ones, and give these supports to which the tendrils can cling. They never attain majestic proportions.

There are other thunbergias that need the stove temperature, notably *T. erecta* and *T. grandiflora*.

WARPURIA CLANDESTINA. A collector's plant that needs similar treatment. It makes a striking pot specimen of glistening pale green leaves with black bands down the centres.

Monstera deliciosa. The inflorescence of the Swiss cheese plant, the house plant whose large, much-cut leaves are more familiar.

◀ *Hedychium gardnerianum*. Ginger lily. Herbaceous plant bearing fragrant lemon-yellow flowers in summer.

Flowers are unimportant, and should not be allowed to form.

The family is rather confusing, but there are species under other generic names including: *Barleria*, *Geissomeria*, *Jacobinia*, *Justicia*, *Peristrophe*, *Phlogacanthus*, *Sanchezia*, *Schaueria* and *Whitfieldia*.

They can generally be considered to respond to the treatment needed for aphelandras.

◀*Medinilla magnifica.* Evergreen shrub bearing rosy pink flowers in May.

CHAPTER 5

The Apocynads (Apocynaceae)

THE family comprises varied forms of evergreen shrubs and climbers, some of which are very useful if there is room enough for them in the stove or warm greenhouse. Their chief quality is to produce handsome flowers for many months each season.

Probably the most renowned is *Nerium oleander*, with its handsome stems and foliage and lovely blossoms, as well as its poisonous content, should anyone be silly enough to eat any part. This and another close relative, the *Acokanthera*, are best grown in the cool greenhouse.

Allamanda

These are climbing plants that produce trumpet-shaped yellow blossoms from late spring until late autumn.

The usual way to grow them is to plant in tubs or large pots, or in the greenhouse border, the standard John Innes potting compost suiting them specially well. One important thing is to make the compost very firm when potting, and always to keep roots as restricted as possible without actual starvation.

In other matters they like the normal stove conditions, i.e. heat and moisture in summer, and warmth and partial dryness in winter. They are trimmed back after flowering finishes, in late autumn. In spring the new shoots are either used as cuttings, or the plants are allowed to develop and, as they grow, are trained to wires near the glass.

The flowers arrive at the ends of the new shoots, like plumbago. It pays to pinch back straggly shoots so that the side-shoots will make a bushy appearance. In fact, if pinched back enough, a compact plant will result.

ALLAMANDA CATHARTICA. Leaves are in whorls, and smooth. Flowers are bright yellow with white centres.

A. C. HENDERSONII. Has orange-yellow flowers with a white throat.

A. C. NOBILIS. Clear yellow flowers without marks.

A. C. SCHOTTII. Yellow flowers with brown throats. They are larger than the species, but the plant is less floriferous.

A. × CHELSONII. This hybrid has large, deeply coloured flowers that arrive profusely. It is the best one for cutting.

Dipladenia

These are the most beautiful and most challenging of the family. They are moderate in growth but, like the allamandas, are best when they have room to climb. They can be trained to wire trellises close to the roof, and only need shading from the sun during the hottest hours. First of all, the cuttings are taken during August and rooted in the very warm propagator. The rooting and potting compost should be of peat and sand, the lumpier the peat the better. At first small pots are used. Then they are put into larger ones, and the twining tops are trained to a temporary framework of bamboos, the base inserted in the pot. Ultimately, they are put into larger sizes, up to 15 in. or more, and the twining shoots are trained up permanent trellises. Or they can be planted in a peat border under the staging. However, drainage must be assured. They resent sourness and wetness. The temperature should be constant throughout the year, for they are evergreen. They start to flower when 18 in. high and continue unabated as they grow upwards and outwards. In November, they are trimmed back to the bare main stems. After a brief rest, new shoots and flowers start to come again. Liquid fertiliser is good for them.

The flowers are excellent for cutting and for putting into bouquets. The chief reason for growing, however, is the marvellous continuous indoor display.

DIPLADENIA × AMABILIS. This hybrid is considered to be the best one in cultivation, both for foliage and the flowers, which are rosy-crimson in colour, and fully 5 in. across. They are borne in clusters and are remindful of small petunias, but of firmer texture, and remain fresh for a long time in water when cut. A plant will attain 10 ft. and bear flowers continuously from late spring until early winter.

D. × BREARLEYANA. The flowers are very large, and open three or four on a bunch at a time. They are deep velvety reddish crimson. They satisfy those who appreciate grandness.

D. CRASSINODA. Is a true species from Brazil, much more slender in habit than the foregoing, and has smaller, shiny leaves. Rose-coloured flowers have yellow throats, and are produced most generously through the long season.

There are many more species and hybrids.

* * *

BEAUMONTIA GRANDIFLORA. The Nepal trumpet flower is a twining climber whose young shoots are pink and which has interesting leaves, but the chief attraction is the large white flowers tinted green at the base.

CATHARANTHUS ROSEUS (syn. *Vinca rosea*). This is the easiest greenhouse species, and flourishes in the stove as freely as the periwinkles (*Vinca major* and *V. minor*) do in the open garden. You can take cuttings any time and grow them in pots of ordinary compost. You can pinch back growing shoots to cause bushiness, but it should be remembered that the flowers are at the terminal points. If left, they produce these continuously from spring until autumn. You can trim back the old plants each winter if you prefer not to raise fresh stocks. In addition to the rose-coloured species, there is a white form.

ECHITES RUBROVENOSA. Evergreen needing identical treatment to dipladenias. The chief attraction is the exceptional emerald-green foliage netted red and gold.

ERVATAMIA DIVARICATA (syn. *Tabernaemontana coronaria*). This has several common names, such as Adam's apple, board-leaved rosebay and Nero's crown. Has lovely white flowers after the manner of oleanders, which are strongly scented by night. A fully-grown tree produces two crops each year. It is evergreen and attains 8 ft. Can be increased by cuttings taken during spring. Ordinary John Innes potting compost is suitable, and fully grown trees should be fed regularly with liquid manure. There is a double-flowered variety that was once popular.

PLUMERIA RUBRA. The frangipani or Indian jasmine makes a tree up to 20 ft. high, and bears fragrant flowers in various colours according to variety. They are waxy and funnel-shaped and come in clusters. Not a good greenhouse subject.

PRESTONIA VENOSA. A twining shrub with leaves containing crimson veins. It bears greenish yellow flowers in loose hanging racemes, during June.

STROPHANTHUS GRANDIFLORUS. This is one of several genera needing stove treatment. It is a climber with flowers more curious than beautiful. They are chrome yellow streaked red, and have tails hanging from the petals.

CHAPTER 6

Aralias (Araliaceae)

ALTHOUGH several genera of this family are very well known as house-plants, very few need the cosiness of the stove or warm greenhouse. They grow well in cooler temperatures; among them are the aralia itself, the fatsia, and a hybrid between the latter and an ivy called *Fatshedera lizei.*

Panax is also an important word in the order, it being not only the genetical name, but also forming part of others, such as *Echinopanax*, *Notheopanax*, *Oreopanax* and *Pseudopanax*. The best known species, *Panax quinquefolius*, known in America as Ginseng, has tuberous roots used for medicinal purposes; the name being derived from the word 'panacea'.

The following species are examples of those that need warm greenhouse treatment. John Innes potting mixture, to which is added moss peat to a third of its bulk, suits them. In other respects the usual treatments regarding temperature, watering and syringing are good. Propagation is easy, but a warm place is necessary for cuttings, or eyes. The plants are inclined to grow leggy. Cut them down hard, and then use the best of the new shoots growing from the top bunch as the easiest way to increase stocks. Some of the variegated varieties are grafted on to plain roots. But this is hardly necessary, for practically all the family are far from difficult to increase, or grow.

During recent times the names have become a bit mixed and involved and what were once aralias are now something else. Among the chief warmth-loving species are:

Dizygotheca

DIZYGOTHECA ELEGANTISSIMA (syn. *Aralia elegantissima*). This makes a handsome small tree for a pot and has a slender stem, at the top of which are elegant compound leaves mottled whitish, on long stalks, the several leaflets drooping attractively.

D. KERCHOVEANA (syn. *Aralia kerchoveana*). The leaves of this are more fan-shaped, in the form of several leaflets of a bright green colour with pale midribs. They are more undulating than the preceding. Both species thrive in a warm greenhouse, and flourish in the stove.

D. VEITCHII (syn. *Aralia veitchii*). This is rather taller, with a slender stem, their stalks and digitate leaves with up to 10 leaflets, the colour being glossy green above and dull red underneath. Quite a handsome plant.

Polyscias

POLYSCIAS BALFOURIANA (syn. *Aralia balfouriana, Panax balfourii*). A bushy shrub with three-lobed leaves forming a kidney-shaped whole. They are leathery and quite attractive.

P. GUILFOYLEI (syn. *Aralia guilfoylei*). Coffee Tree. Has an erect stem and shrubby habit. Will grow to 12 ft. in a heated greenhouse. The leaves have several lobes about 2 or 3 in. long, evenly margined white.

P. G. MONSTROSA (syn. *Aralia monstrosa*). Has pendant leaves with very irregular lobes that are margined white and splashed grey.

P. PANICULATA (syn. *Terminalia elegans, Panax paniculata*). Has smooth leaves with one large terminal some 8 in. long, together with several side leaflets. This is sometimes grafted on to stronger growing aralias, as young plants on their own roots take a while to grow. Colour is deep green, with a red midrib, and a red netting throughout the rest. It prefers shade to bright sunlight.

* * *

SCHIADOPHYLLUM CONICUM. A shrub with leathery leaves containing up to a dozen leaflets. Grown mainly for the large glossy leaves. Needs stove conditions and normal treatments.

TREVESIA PALMATA (syn. *Gastonia palmata*). Has lobed, palm-like leaves quite 2 ft. across. This shrub is prickly and some parts are hairy. It has green-white flowers in long panicles during spring.

T. SANDERI. The impressive leaves of this one are circular, fully 2 ft. across and deeply lobed.

CHAPTER 7

The Aroids (Araceae)

THE familiar wilding of this family is the cuckoo-pint that grows on shady banks and by hedgerows. The best known florist's flower is the arum lily, seen at funerals and weddings. This needs cool house treatment, rather than stove or warm greenhouse. There are genera covering hundreds of species of the family that do need extra warmth, and often swamp moisture. They include some that climb, such as the Swiss cheese plant. Quite a number make excellent houseplants, because they are able to withstand shade. They are not difficult to propagate or to maintain.

According to their season, so they need periods of rest. While resting, they need little water—in fact, some want almost complete dryness. On the other hand, the evergreens never really go to sleep and always need some moisture. They tell you themselves. When flowers cease to come and the foliage fades, water should be withheld and only given when the compost becomes almost dry. But as long as there is an urge to grow they must be kept moist. If they completely fade off, keep them practically dry until early spring, repot and then give them just a little water until they show signs of growing again. Then syringe them often, and increase the water supply.

The compost should be rich and yet porous for, although they like plenty of water while growing, it must never be allowed to get sour or the roots quickly rot. Old formulas invariably gave cow manure as a necessary ingredient. If it is possible to get that commodity in a dried state, it can be added to form a sixth part of the bulk, but in the ordinary way the John Innes potting compost is eminently suitable. However, the loam ingredient can be reduced by adding a quarter the amount of sedge or Irish peat, and a quarter part sharp, gritty sand.

When potting time comes, usually in February or March, only use sizes that contain the roots comfortably. Keep moderately moist at first. If any new stock is wanted, take away new shoots, each with a piece of tuber, or rhizome complete, and put singly into small pots using a mixture of half peat and half gritty sand. If kept in a shady, moist, warm place, fresh roots develop and the resulting plants will soon need larger pots and the more substantial compost already described. The group which follows contains the aroids, which are grown for the beauty of their foliage.

Aglaonema

These are far from the most interesting of the genera, but are much grown on account of their ability to stand living-room conditions. They are perpetual; in fact, a common name in America is Chinese evergreen. The foliage is the chief attraction.

AGLAONEMA COMMUTATUM. Has rather broad, oval leaves about 9 in. long, deep green in colour, spotted paler green and white. Grows about 2 ft. high. The spathe is white.

A. COSTATUM. Has broad, shining green leaves about 5 in. long, that are spotted white. This is a dwarf plant with a largish spadix and green spathe.

A. PICTUM. Has deep green leaves marbled grey and white, is moderately dwarf, and has a white spike with creamy yellow sheath.

These are typical of many species and some varieties. Practically all grow well in a warm greenhouse.

Alocasia

This genus contains many species and hybrids of diverse forms and colouring that are not easy to grow. They need much the same treatment as the other brilliantly variegated aroids, but more drainage should be put into the pots throughout their lives, and they need the warmest part of the stove, where they get light but not direct sunshine. They grow very quickly and should be given liquid manure or fertiliser at regular intervals throughout the lightest months of the year.

ALOCASIA CUPREA. Has tuberous roots, leaves are shield-shaped, about a foot long, and of a purplish metallic tint with violet undersides. Will make a plant 6 ft. high and across under good cultivation.

A. LOWII. Has arrow-shaped leaves over a foot long, that are dull green with white ribs, and violet on the underside.

A. L. VETCHII. Has long, arrow-shaped leaves with ivory-white veins, and a slatey metallic sheen on top. They have purple undersides.

A. × SEDENII. A hybrid between *A. lowii* and *A. cuprea* with leaves about 15 in. long, metallic dark green on the top surface and violet beneath.

There are a few dozen more, some now changed to *Colocasia*, notably:

COLOCASIA AFFINIS JENNINGSII (syn. *Alocasia jenningsii*). A small growing plant from India with leaves about 8 in. long, rich green and blotched with black or chocolate spots.

C. ANTIQUORUM ILLUSTRIS (syn. *Alocasia illustris*). Somewhat similar but larger with deep green leaves spotted olive green.

Anthurium (see also p. 77)

In addition to anthuriums primarily grown for flowers, there are many species and varieties favoured for their foliage. They need similar conditions, and may be increased in the same way. They can also be raised from suckers at any time. They love heat and moisture and must never be dried off, although they have resting periods when water should be given sparingly.

ANTHURIUM BROWNII. Has short stems with leaves over 2 ft. long and 15 in. across. They are leathery olive green, and are more important than the green and purple flowers.

A. CRYSTALLINUM FOLIIS VARIEGATIS. Has beautiful large leaves over a foot long, that are marbled ivory with yellowy shadings.

A. TETRAGONA. Leaves are up to 4 ft. long, with shining, deep green surfaces on top and paler hues underneath.

A. VEITCHII. Its leaves are over a yard long, like polished green leather. Flowers are insignificant.

Note. There are also climbing anthuriums, but hardly serious rivals with other genera for places in the choicest greenhouses.

Caladium

Caladiums should be kept practically dry in winter, whereas the evergreen or almost evergreen species of the other arrow-head tropical swamp aroids must not be allowed to get drier than moist. To reiterate an earlier statement, the best guide is to let the plants tell you. If the tops completely die, give them a rest until spring. If they remain green, keep them moist throughout.

Between the two world wars, caladiums were regularly shown at Chelsea and were much grown in large gardens. They are probably the easiest and most satisfactory of ornamental aroids. They flourish without much trouble in ordinary compost and prefer the stove temperatures. Most of them definitely go to complete rest in winter, just as do tuberous begonias and gloxinias, and their tubers should be kept dry in the pots until late February or March, when they are started into life again.

Bought tubers, and those tipped out of their pots in early spring, are generally started into growth by placing them in a box of damp moss or sedge peat placed in a temperature of 21°C (70°F). Once they start to grow they are then put singly into 4-in. size pots or whatever they can be held in just comfortably, and kept growing by syringing, so creating moist warmth (a cucumber house is a good place).

They are, most times, increased by dividing the tubers when turned out of the pots at the end of the resting period, but the best method is to get the tubers growing first and then, when there are new growths about 5 in. high, take these off with some rootlets attached. Plunged into the closeness of the propagating frame, they soon develop into nice specimens.

CALADIUM BICOLOR. This is the name given to the traditional parent of the hundreds of cultivars that have been raised through the years. We are told that it came from Brazil, but it seems no one knows for sure whether they have seen the true species. However, the divergence of colours and marking of the leaves are indescribably remarkable. Among them are green, white, grey, rose, red, violet, in a thousand shades, often many in one leaf.

C. B. CHANTINII. This has green, red and white colourings, either as the basic colour, or as veins, or spots, in the leaf make-up, while the long substantial stalk is violet.

C. PICTURATUM ADAMANTINUM. White-spotted veins on handsome leaves, with pink and brown markings on stems.

C. SCHOMBURGHII PICTUM. Has red veins in the leaves that are spotted white and red.

Colocasia

see Alocasia

Dieffenbachia

These evergreens are most suitable for the warm greenhouse. They make a very bold display, some species attaining 6 ft. Stems are thick and fleshy, while the handsome, broad, variegated or marked leaves are sometimes each 2 ft. long and a foot across. Blue flowers arrive in the autumn.

All the species contain a poisonous sap which, if any is taken in the mouth, will cause swelling and speechlessness for several days. The common names of dumb cane or dumb plant were given on that account.

In most respects cultivation is much the same as for the other ornamental aroids. But being evergreen, the roots should never be allowed to become entirely dry.

They are quick growers, and the best procedure is to raise fresh stocks every year. Cuttings can be rooted during early spring for the autumn display, or they can be taken later to obtain specimens for the following season. Any part of a stem can be used, provided there is a growing eye upon it. This should be cut out and planted in a small pot with the top just covered as for grape vines, and placed in a warm propagator. The best potting compost is John Innes, with additional peat added.

DIEFFENBACHIA BOWMANNII. Has large, emerald-green leaves with prominent, lighter green markings.

D. CHELSONII. Has shot silk, dark green leaves decorated with grey veining and yellowish spots.

D. EBURNEA. Smaller than *D. bowmannii.* Leaves are beautifully marked or dotted white and whitish pink.

D. PICTA. Green leaves spotted white and paler green.

D. P. BAUSEI. Paler than above with white spots on leaves and white stalks.

D. SEGUINE. This is the most favoured species, the original having deep green leaves speckled with white spots that are almost transparent. They attain 6 ft., and after a while grow leggy. Some gardeners head them back, to cause bushiness. Constant renewal is the best policy.

D. S. BARRAQUINIANA. Stalks are white, and green leaves are blotched. They are smaller than the type.

D. S. LITURATA. The shiny green leaves have grey arteries and markings.

There are many others.

Homalomena

This is another genus with lovely leaves that need either warm or stove conditions. The treatment is normal. Some go completely dry in winter and should be treated accordingly. Two typical species are:

HOMALOMENA PELTATA. Heart-shaped leaves about 2 ft. long, of a fleshy green colour. Height is about 3 ft.

H. RUBESCENS. Dark green leaves more arrow-shaped than above, that are deep green above and purple underneath.

We now come to the aroids that are grown more for flowers than foliage, although the latter is very attractive. We must omit many species that have undesirable qualities, such as foetidness, or which are far from picturesque, but mention must be made of the 'monarch of the east', although it is not a stove or warm house subject. *Sauromatum guttatum* is offered by bulb merchants as a 'miracle plant', for, if the tuber is kept in a warm place in spring, it will produce a flower without being given soil or water. Then, if it is planted outdoors in a moist position, it will form leaves and recover sufficiently to be dug up again in October, and stored in a dryish cool place until February comes again. It will be likely to do this for years.

Anthurium

This is an exciting genus. The flamingo plants are a good example. These are the anthuriums grown for the amusingly queer flowers, with piggy-tail spadices that arrive spasmodically through the seasons. There are other species grown purely for the foliage (see p. 74). Of all the aroids, there are probably more species within this genus than of any other. There are surely more varieties and cultivars.

Their cultivation is not very difficult. If the normal stove or warm house conditions are given, they get the moisture they require at the right times of year. They are best grown in compost that drains, but holds moisture well. Irish moss peat added to John Innes potting compost will do this very well. Some advocate adding sphagnum, but this is only necessary when they spend most of their time in living-rooms.

We can propagate them by carefully sorting out the roots during February, and taking from them the apparent crowns, that can then be potted into a mixture of peat and sand and placed in the propagator. It is best to have them with rootlets attached. Once they grow they do so quickly, and can soon be moved into larger-sized pots.

Although it is usual to see rather puny specimens with few flowers, it is possible to grow them to a size when they bear a score or more, well above the pleasing leaves that overlap like thatch. Always remember, they like moisture and warmth. And they benefit from a little liquid fertiliser given frequently while growing freely.

Some flowering species include:

ANTHURIUM ANDREANUM. A beautiful species, of which there are many variations. They have large spathes, or shields, heart-shaped and corrugated, bright scarlet or deep orange, looking as though varnished. The spadix is white at the base, with a yellow tip. It droops somewhat, adding to the colour contrast and shape. There are so many varieties or cultivars of this species that it is invidious to name any of them. They range from those having white spathes to those of yellow, pink and ivory. Heights and sizes vary considerably.

A. SCHERZERIANUM. Differs from the foregoing in length of leaf and form of spathe, although its varieties are numerous. Colours of spathes and spadices vary beyond description. Some are brilliant scarlet. Others can be purple, crimson, red, rose, salmon, yellow, white or green. Some are dotted or spotted another colour, while others can be splotched with varying shades of the basic hue.

A. S. WARDIANUM. This is a variety that has broad scarlet spathes fully 6 in. long, and is regarded highly among specialists.

It should be noted that the variants of *A. scherzerianum* are easier to manage than those of *A. andreanum*. The latter will flourish in a warm greenhouse.

Monstera

These are closely related to the philodendrons. They need more room, and specimens are often seen in hotel and hospital reception rooms. They are grown easily in the normal stove or warm greenhouse.

Monsteras will flourish in comparatively small receptacles, as long as the powerful aerial roots are left to absorb goodness from the atmosphere or cling to other things for support. They have been known to penetrate carpets if they can find goodness in them.

The large, elephant-ear leaves are much indented with patterned gaps, so that they can cope with hurricane attacks in their native jungle giving a decorative effect of character rather than of beauty. Other cultural needs are as for philodendrons.

MONSTERA DELICIOSA. Swiss cheese plant. Mexican bread fruit. M. PERTUSA. These two species are so much alike that they are often mistaken for each other. According to the botanists the chief differences are in the vigour of growth and size of leaves. *M. deliciosa* will, if given root and top room, attain 20 ft. and produce serrated leaves fully 18 in. across. It probably produces the most and best fruit that has a good flavour. This species can be kept to a respectable size by retaining the roots in a largish pot. *M. pertusa* is regarded as a smaller version of the above with more rounded leaves that spread merely to about 10 in. However, the species cannot be entirely defined by form, it being variable. There is a monstera often sold in florists under the title *M. deliciosa* 'Borsigiana' and also *M. pertusa* 'Borsigiana', the latter being the more correct. It is one of the best forms to grow.

Philodendron

These are practically all climbers that clamber over other vegetation. They produce aerial roots which, if they can cling to other stems and boughs, absorb moisture, giving them more vigour. So, one of the items in their cultivation is to insert pieces of cork bark in large pots to form a support to which they can cling. It keeps them moist.

However, if the aerial roots are cut away, the plants are slower in development and can be grown as dwarf bush specimens. For all that, the best way is to train them up trellises, fixed where there is some shade, such as the back wall of a lean-to greenhouse. We must remember that they get the shade of their hosts in the jungles. They will thrive in a warm or a stove greenhouse. Many species succeed in living-rooms, being tolerant of the warm dryish atmosphere and the paucity of light.

Because of their adventitious rooting, they increase easily. Merely cut a stem into three or four pieces, each a few inches long, and stick them into a compost mixture of sand and peat; they are best grown in a no-soil compost. They root quickly when kept warm and moist.

PHILODENDRON ANDREANUM. It has pendulous leaves, typically like those of caladiums and arums, but up to a yard long, with a surface of dark green that scintillates a shot-golden sheen.

P. BIPINNATIFIDUM. The leaves of this are heart-shaped, but much serrated when fully developed and over a foot across. They make 'parlour palms'. Tolerates cooler conditions than most species.

P. ERUBESCENS. Its leaves are more heart-shaped than the foregoing, and much shorter, the leaf stems also being much reduced. Shining green leaves have a deeper underside.

P. LEICHTLINII. Remarkable on account of the heart-shaped leaves, studded with lengthy oval holes as though eaten by caterpillars.

P. VERRUCOSUM. Satiny green leaves banded with maroon, having reddish stems decorated with fine hairs.

Note. There are many more lovely philodendrons, and numbers are increasing on account of their suitability as house-plants.

Scindapsus

This is a genus of climbers, some vigorous; others are trailers, being to some extent controlled by taking cuttings from various parts of the old plants, according to their vigour. In most respects the cultivation is the same as for philodendrons. Most flourish in the warm house and are suitable for living-rooms, being extensively grown for this purpose in American and Continental homes. A peaty mixture is preferred to the standard John Innes compost.

SCINDAPSIS AUREUS. Also known as the ivy arum, or Ceylon pepper, this makes an impressive climber, with heart-shaped leaves and short aerial suckers. It resembles in some ways variegated ivy, the marbled yellowing effect being similar.

S. PICTUS ARGYRAEUS. This is a twining shrub with a bluish hue, dotted or marked with silver. There are cultivars offered by florists.

Spathiphyllum

These are evergreen plants, sometimes grown as house-plants, which need the same treatment as anthuriums. They have erect spathes, while the spadices stand up somewhat in the manner of plantains. In fact, they remind me of emaciated hostas.

SPATHIPHYLLUM FLORIBUNDUM. Has leaves about 6 in. long. Spathes are white, and larger. The spadix matches.

S. WALLISII. Leaves are larger. The flower is pure white. Spathe is insignificantly green.

Syngonium

These are climbers of less vigour than others previously mentioned, but needing much the same treatment. They can be grown in pots, and trained up canes. Although revelling in stove conditions, they tolerate room atmospheres and temperatures, and for that reason are generally included in lists of house-plants.

SYNGONIUM AURITUM. The common name of five fingers refers to the leaf formation, it being usual to have two pairs of smaller leaves at the base of the main one which is green, and about 6 in. long.

S. PODOPHYLLUM ALBOLINEATUM. Has arrow-head-shaped leaves that are deeply lobed, while midribs are marbled with white.

S. VELLOZIANUM. In this case only one pair of smaller leaves are formed at the base of the main one, and it is therefore called 'three fingers'. There are several varieties.

Zantedeschia

In addition to the ordinary arum lilies that grow best in cool greenhouses, there are several species with coloured spathes that need warmer conditions. The warm greenhouse is preferable to the stove.

When the dry roots are bought in winter they are immediately put into shallow boxes of peat, placing these in the warm greenhouse to keep them warm and dryish until February. They are then potted into John Innes potting mixture, to which dried cow manure or peat has been added, and water given. Once they grow, water more often and syringe from time to time. They eventually produce flowers in early summer. Once these and the foliage begin to fade, reduce the watering and put them in a cold frame until October. After that bring them into the warm greenhouse to repeat the treatment. If more plants are wanted they are obtained by taking suckers complete with rootlets.

ZANTEDESCHIA ANGUSTILOBA. These have until recently been known as *Richardia pentlandii* and grown by skilled gardeners, who seem to appreciate their brash beauty much more than do their employers. The deep, glaring mustard-yellow spathes have large purple blotches at the base, and are the main attraction, although leaves are typically arrow-headed.

Z. ELLIOTTIANA. The yellow spathes of these are long and clear, while the leaves have transparent blotches all over them.

Z. REHMANNI. Not as robust as the above. The spathes are pinkish, the shade depending upon the variety.

CHAPTER 8

The Asclepiads (Asclepiadaceae)

THERE are nearly 2,000 species in this family, most of which may be disregarded in this book. Two definite types are grown in greenhouses. One is the cactus-like stapelia that comes from South Africa and thrives in cool temperatures, as do several other related genera. But for another reason, they are not recommended for the stove or warm greenhouse—they stink. The common name is carrion flower, because they possess an inbuilt factory that emits a smell to attract the flies that are necessary for fertilisation. They are also called starfish flowers on account of the most wonderful, if somewhat weird, blossoms. More desirable are the twining and clinging plants that thrive in the warm house rather than the stove. If grown in the higher temperature, they have not the same quality, although propagation is easier if cuttings are rooted in a very warm propagator.

Asclepia

ASCLEPIA CURASSAVICA. The blood flower, or milkweed, is one of the very few non-climbing species that needs a warm house, excluding any of the stapelia types. It forms a squat shrub if pruned down each winter. The tips of the growing shoots should also be nipped to cause bushiness. Deep purplish flowers arrive throughout summer. In other respects the normal stove greenhouse treatment will please it.

Hoya

This twining and climbing genus is quite popular. The flowers hang down in neat umbels, and the best position is where they can be looked up to. One good place is along the cross stays of the house, but they will also cling to pieces of wood or bark, forming aerial roots as and when needed.

The roots proper should be in large pots, or for specimens in borders. Most suitable compost is the John Innes potting mixture, increased by adding one-third part sedge or moss peat.

Propagation is by cuttings of shoots produced during the previous year, which readily root in a propagator. Some growers layer the branches into sandy peat, after removing the lovely leathery leaves that are in the way.

Water should be given fairly frequently in summer and syringing is appreciated. But in autumn the plants go to rest quite early and then only need enough to stop the leaves from flagging.

It is interesting to know that the name *Hoya* was given to the genus in honour of Thomas Hoy, a head gardener to the Duke of Northumberland, at Syon Park, Brentford, Middlesex, where the glasshouses are still a feature.

HOYA BELLA. This is a connoisseur's version of the genus, growing not more than 2 ft. high, and drooping over to form a compact bunch of smallish, neat leaves. Small wax flowers are white, with brighter coloured centres. This species needs a stove temperature.

H. CARNOSA. Commonly known as the honey plant or wax plant, this is the most popular species of five or six dozen, and climbs up to 8 ft. or more, adhering to anything that has a suitable surface. A piece of oak bark fastened to a stick and inserted in a pot will please it. Flowers are white with pink centres, and hang in clusters like tiny inverted parachutes. The leaves are fleshy and handsome and always look pleasing. The plant is best given light, but will tolerate some shade. Any pruning should be carried out in February, but the spurs upon which the flowers come should not be removed from the remaining stems, or the next crop of flowers are spoiled. The best position is in the warm greenhouse, rather than the stove.

H. GLOBULOSA. A more refined version of *H. carnosa* with creamy flowers in ball-shaped umbels.

For collectors there are many other good species, differing only in detail from *H. carnosa*. One firm lists a dozen distinct sorts.

Stephanotis

This is undoubtedly the doyen of the climbing and twining asclepiads. In contrast to the stapelias, its fragrance is renowned for its marvellous sweetness. Chief points in its cultivation are to give plenty of root room, and not to repot more than necessary. A border in the greenhouse is better than a pot if it can be provided. The compost can be either peat or loam. I prefer John Innes, to which is added a third of moss or sedge peat.

Propagation is by cuttings taken from the shoots of the season before. They can be reduced to two joints, and several can be placed in a pot of half sand and half peat and placed in a temperature of 16°C (60°F). If potted in winter and kept moderately moist, they will callus over, to root eventually in spring. Alternatively, you can take off new shoots in spring, each with a heel, and keep moist and warm in the propagator. Once rooted, they are potted up singly, and will grow steadily.

Care must be taken to tie the shoots to a trellis before they become entangled. When about 2 ft. high, nip out the tops to cause more side-growths, and likewise train these to the trellis or wires.

This plant likes less moisture than usual, and should be given sunshine to the point of not scorching. Keep the roots moderately watered, and syringe the tops in the afternoons only during the growing season. Encourage plenty of growth in their early years, but as the plants grow older, shorten back the shoots directly they have finished flowering. They like ventilation during warm weather. Old plants need feeding.

STEPHANOTIS FLORIBUNDA. This is the main species, with the dark, evergreen, thick oval leaves and vine-like stems that will grow up to 10 ft. The waxen flowers, similar in shape to tobacco blossoms, are pure white throughout, and intensely fragrant, giving it its common names of the clustered Wax Flower and Madagascar Jasmine. They arrive continuously on established plants through late spring and summer, being less free on younger specimens. They sometimes bear fruit that resemble plums.

Other genera in the family that climb and need similar treatment to the hoyas are: *Arauja*; *Marsdenia*; *Oxypetalum*; *Wattakaka*.

CHAPTER 9

Begonias (Begoniaceae)

OF all indoor flowering genera dealt with in this book, it is most likely that begonias exceed and excel them all. Yet they are given moderate importance here because it would need a book of considerable size to tell all that could be written about them. During a lifetime's experience, all sections have come under my care, and I realise that most are for the specialist, rather than for inclusion in a general stove or plant house.

Very broadly, there are three types; the tuberous, the fibrous-rooted and the ornamental-leaved. These, in turn, may be sub-divided, as in the case of the tuberous ones, into the large-flowered, the multiflorus and pendulous.

The *tuberous* begonias have been intensively bred from several original species—including *Begonia bertinii*, *B. baumannii*, *B. bicolor*, *B. boliviensis*, *B. cyclophylla*, *B. davisii*, *B. dregei*, *B. evansiana*, *B. froebelii*, *B. fulgens*, *B. gracilis*, *B. modica*, *B. pearcei* and *B. veitchii*. I understand that some of the original species are possibly obtainable, but those mainly grown are of the limitless cultivars.

Stocks are started by bought tubers, or grown from seed sown in the early part of the year (the stove or heated greenhouse being convenient for this) before they go to a cooler house, or are planted in the open garden. How to cultivate them is described by almost every good seed catalogue, or instructions are issued by specialist firms like Blackmore and Langdon, of Bath.

Next are the *fibrous-rooted* varieties that include the very popular wax begonia, most times listed as *B. semperflorens*. Most seed lists have several cultivars, including in recent times the F_1 hybrids. They are best treated as annual bedding plants which can be started earlier than usual by placing the seedpans in an available greenhouse.

In another fibrous-rooted section are many cultivars we gardeners have specialised in, because they bear flowers in winter, and are most often referred to as 'Gloire de Lorraine' begonias. Actually, the name covers a considerable number of hybrids and cultivars, derived in the first place from a famed species, *B. socotrana*, which has a bulbous root and roundish leaves, while the flowers are rose-coloured and some 2 in. across. One of the other parents is *B. dregei*, which comes from South Africa, and grows about 2 ft. high, with leaves more like those of geraniums; it has smallish white flowers.

They are definitely stove plants, but different to the general collection in that they are growing and flowering while other genera are resting, and they need the rest when others are awakening from the winter torpor. They should be watered moderately, while the growing shoots must be trained and tied to canes or, preferably, thin peeled sticks. A chief point is always to keep them warm, and give plenty of light and shade from fierce sunshine, which is rare in winter, in Britain in any case. After flowering, they are cut down and rested, but water is not entirely withheld. About April, they start sending up shoots again when it is time to repot into suitable sized pots. Good, well-drained, rather rich compost is suitable. The usual John Innes potting, lightened by adding a third part of moss peat, is good. To add to the richness, some dried cow manure rubbed down to small lumps and mixed in, is excellent.

As growth develops, the shoots are pinched, or tied to supports to make nice shapely plants, and the tips can be rooted in a sandy-peat mixture in a warm propagator without much difficulty. Old plants are grown on, the cuttings are also grown on, and in due time rested, to make nice plants the next season.

They can be propagated from leaves. A mature leaf is picked off a plant, and its ribs are severed in several places from the underside. It is then laid on a pot or pan of moist, soft sand, with the top of the leaf uppermost, and held down by a peg, or a little sand placed upon it (see Fig. 6). Young plants form at the cuts, and are lifted carefully for potting off singly in a light compost.

There are many other begonia species, mostly of a shrubby character, which make good, permanent specimen plants in warm houses. Foliage is attractive. Habit is desirable and flowers are an additional feature, although not to be compared with the brilliant individuals of the tuberous cultivars. Some even climb, and can be trained to trellises on walls, much like outdoor wall shrubs.

They are not fussy about compost, as long as it is rather rich and porous. They can be propagated from stem or leaf cuttings, or have the roots, sometimes rhizomes, divided into pieces. They can be encouraged to grow bushy and tall, or can be reduced

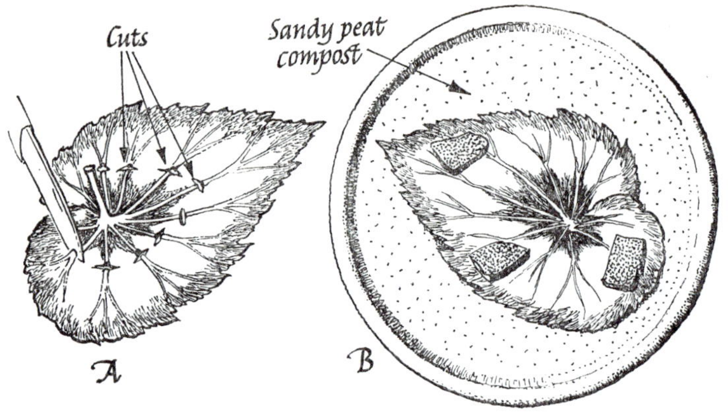

Fig. 6. Leaf cuttings of Begonia.
A—Cut through the ribs at the back.
B—Turn over and place on damp sandy compost in a pan, insert cut stem and pin other parts down to keep flat, or place two or three pieces of pot crocks to hold down (see text).

from time to time. I like to damp down among the pots on the staging, but never directly syringe the foliage. Semi-shade is best for them, and they liven up a fernery, should this be the form of warm greenhouse preferred.

BEGONIA COCCINEA. This will attain 4 ft. The stems are straight and smooth, while leaves are oblique, narrowing down to a rounded point at the base, suggestive of a wing—hence the common name, angel wing. Some varieties have leaves 6 in. long that are silver spotted. Flowers are smallish, in clusters, of a coral-red. They arrive through spring and summer.

B. × FEASTII. The beefsteak begonia has almost round leaves, with thick white hairs on the margins. They are green uppermost and red underneath. Flowers are light pink. This is an easy plant to grow.

B. HAAGEANA. This is a very popular stove or warm greenhouse subject that will grow about 2½ ft. high. The leaves, of varied sizes up to 6 in. long, are red-veined on a pellucid green surface, with a purplish colour underneath. Flat heads of smallish rose-pink flowers appear at all times of the year.

B. HERACLEIFOLIA. The star begonia is a tremendous species with leaves that vary in size up to 10 in. across, down to much smaller ones that are stained silver, red and almost black. The pale pink flowers are borne on yard long stalks. There are several varieties and cultivars.

B. METALLICA. This grows into a handsome plant 4 ft. high. The shining olive-green leaves have burnished purple veins, and are reddish underneath. The clusters of light pink flowers are fairly large and appear during summer and autumn.

B. PHYLLOMANIACA. This bears pink flowers during winter, but is more renowned for the numerous tiny plantlets produced upon the stems, stalks and leaves, which can be rooted and grown separately if any are wanted.

These are merely a cross-section of the shrubby species that make good specimen plants as grown in private gardens, municipal parks and specialist nurseries and florists.

The chief of the *ornamental-leaved* begonias is *Begonia rex*, but there are many other species, and untold varieties. They are all recognisable by the unmistakable lopsided shapes of the leaves, after the style of an artist's palette, the stalk being where the thumb normally goes. They are large, although variable, and most difficult to describe, for they are often wrinkled, toothed and hairy, and are covered with an array of many metallic colours in zones, or blotches in infinite variety.

These are more tender than the shrubby kinds, and prefer shade to sunshine. The best place is the warm greenhouse, where they are shaded by other plants, the average height being 9 in., although some will attain 2 ft.

They love moisture, but not wetness; drainage but not dryness; light but not sunshine; comparatively small pots but plenty of food, hence need feeding regularly with liquid fertiliser. Some advocate all-peat composts. These are all right if feeding is carried out later. I still prefer John Innes potting compost, to which is added peat to make the soil more porous, yet moisture retaining.

BEGONIA DAEDALEA. Has short stalks and stems with the same type of leaves as *B. rex*, the chief character being russeting surfaces on the hairy edged leaves.

B. IMPERIALIS. Leaves are about 5 in. long and have a comfortable olive-green surface that is lightened by uneven veining of a brighter hue.

B. MASONIANA. The iron cross begonia has leaves which are rather bristly and have a definite brown pattern that suggested the common name. This is a popular new species.

B. RAJAH. The leaves of these are more round than most and have brown markings between veins, giving a bristly effect.

B. REX. This is the Rex begonia from which most of the many cultivars originated. It grows about 2 ft. high, and has the typical ornamental leaves of metallic green splashed with purple and silver.

Note. Many other species, varieties or cultivars of ornamental begonias, as distinct from tuberous ones, are available from specialist nurseries.

CHAPTER 10

The Bromeliads (Bromeliaceae)

THE pineapple family contains at least a thousand species, of which quite a number make exciting stove plants. By far most are epiphytic by nature, which means that they live perched upon other vegetation, but do not rob their hosts of any goodness; they rely upon the atmosphere and decaying materials for sustenance. They are also xerophytic, which means that they can withstand long periods of drought by staying semi-dormant, simply existing upon the moisture transpired by their hosts, and also taking advantage of the shade from the canopy of foliage overhead.

The most important point to remember is that the roots must be in a potting mixture that drains thoroughly soon after watering. Always put plenty of crocks in the bottom of the pots, which should be comparatively small proportionately to the size of the plants.

A good mixture is 4 parts osmunda fibre, with 2 parts sphagnum moss, a little potting sand, and powdered charcoal. If you live where sphagnum grows, collect and chop it into small pieces, before adding to the other ingredients. Some growers use beech or oak leafmould instead of osmunda fibre. Yet again, there are those who add pine needles.

Some cultivators get lengthy pieces of bark, wrap osmunda or sphagnum around them and bind fairly tightly. Into this the young offsets are inserted and, with occasional sprayings, the roots form, sometimes extending downward under the moss, resulting in good plants in quick time. However, the disadvantage of that is they can get top-heavy, especially when they are of the larger species.

The old gardeners used bracken peat, filled with rough decayed roots and other decayed vegetation, such as the leaves of the trees under which bracken grow. They added sand, charcoal and broken bricks, or pot crocks smashed down to hazel-nut size. Really, any compost will do if it retains moisture, but never becomes sodden, and quickly sheds free water. Do not use John Innes.

Many of the epiphytes are fitted with cups, urns or goblets created by the leaves clustering and overlapping each other at the base, so tightly that water is retained. When these are kept filled, there is little need to worry about the root parts; the spillings, together with the general atmosphere of the greenhouse, being sufficient for them.

The flowers grow from the goblets. After they perish, the main parts of the plants also fade away. From the base of the stems appear new side-growths that will, if left, replace the main parts. But it is usual to take these away, and repot them singly into smallish-sized pots in one of the mixtures just described. If taken too prematurely they do not grow, but just rot. It is best to leave them upon the old plants until they are about one-third of the size of the original. They can then be slipped off by pulling downwards with the fingers or, if too tough, the knife is used to release the parts that insist upon staying attached. They should invariably have incipient rootlets, and be inserted in the new mixture so that these are undamaged. Keep the potting mixture merely moist and sprinkle the tops occasionally, but never overdo it. Sometimes, an old plant with one bold top will produce four or five new plantlets around its base. After these have been taken away, it pays to retain the parent, for other side-growths may also subsequently appear.

In some instances, there is a definite stem between the compost surface and the lowest leaves. Insert in the small pot up to the leaves. In others, where the leaves start at ground level, one or two of them are gently pulled away so that the plantlet can be installed deeply enough.

Aechmea

Among the many genera of the family, I regard the aechmea to be the most valuable and desirable. Most of the species have beautifully marbled and dusted, strap-shaped, recurving leaves edged with spines. They form large deep rosettes, watertight in the middle, that collect the moisture, often being as large as teacups.

AECHMEA CALYCULATA. Is a dwarf species with the usual strap-shaped strongish leaves, while the erect scape produces yellow flowers amid red bracts.

A. DISTICHANTHA. Has an upright rosette of grey-green leaves with pointed ends, decorated with dull red spines. The spikes are clothed with brilliant red bracts, the flowers are purple and have pinkish red sepals.

A. FULGENS. Is not as impressive as *A. rhodocyanea*, although regarded by some connoisseurs as a better species. The broad leaves of a deep green colour flay out like those of the aspidistra. There is the central cup, while the scarlet scape forms into a bottle-cleaner, rather than the duster-shape of *A. rhodocyanea*. The flowers are purplish, and the sepals bright red, rather like berries. The season of flowering is autumn and winter.

A. HYSTRIX. This is rather rare, whereas many of the other species are obtainable. It has spiny leaves and a spike of closely packed flowers of purple colour amid scarlet bracts. The height is quite 2 ft. It arrives in early spring.

A. LUEDDEMANNIANA (syn. *Lamprococcus caerulescens*). Has a rosette about 2 ft. tall, with leaves that are tinged red. The large dense head bears blue flowers, followed by white berries during late autumn.

A. MARIAE-REGINAE. Grows about 2 ft. high. The upper half of the scape bears bright rose-pink bracts, with blue-tipped flowers that fade to pink. The season is late summer.

A. NUDICAULIS (syn. *Bromelia nudicaulis, Aechmea cornui*). Has a rosette of leaves up to 3 ft. long, and a spike of yellowish flowers with red bracts.

A. N. AUREO-ROSEA. Has the usual goblet, the leaves being grey-green, while the flowers and bracts are red.

A. RHODOCYANEA (syn. *A. fasciata, Billbergia rhodocyanea*). Variously known as the air-pine, urn plant, or vase plant, this is my favourite. It forms a strong rosette of succulent leaves of considerable size. From the cup emanates a strong scape, that forms at the top a lovely, glistening, stiff, prickly head the shape of a feather duster. There are bracts of glistening salmon-pink with prickly tips, through which peep Cambridge blue flowers. This scape lasts in good shape for two or three months. The period of flowering varies, but most times is during spring. Best position is the stove while small, the warm greenhouse while developing, the centrally heated room while in flower, and the stove again when the old centre is fading away. Keep from bright sunshine, otherwise provide as much light as possible.

Note. The above are but a few of the best species, the numbers available increasing year by year through the activities of the British Bromeliad Society. Among names appearing in catalogues other than those already mentioned are: *A. benrathi* 'Cardinalis'; *A. blumenavii*; *A. bromeliiflora*; *A. compacta*; *A. cylindrica*; *A. filicaulis*; *A.* 'Foster's Favourite'; *A. gamosepala*; *A. gracilis*; *A. lamarchei*; *A. multicaulis*; *A. multiflorum*; *A. organensis*; *A. pectinata*; *A. recurvata.* There are still more for the collector.

Ananas

To grow commercial varieties of the pineapple (*Ananas comosus*) needs considerable skill. At one time they were cultivated in large estate gardens to the extent that most establishments had their much-heated 'pine house'. They needed a good compost of loam, peat, or leafmould and dung, and were propagated from seed, offset suckers, or the crowns of the fruiting parts cut off to be rooted in a propagator, which they did readily. The procedure included keeping up a considerable heat, and giving them full sunshine. For the latter reason they were not suitable for ordinary stove conditions where light shade is necessary. However, there are some ornamental varieties of the original species that can be grown in the mixed plant house, if fruit is not needed. These are handsome and withstand the dry warmth of centrally-heated rooms quite well.

ANANAS COMOSUS. The pineapple, whose usual height is 3 ft. Flowers are mostly purplish, but are not as lovely as those of other members of the family.

A. C. PORTEANUS. A variety with a broad yellow band down the long dense rosette of leaves. There are several forms.

A. SAGENARIA (syn. *A. bracteatus*). Has three dozen or more leaves about 4 ft. long. Flowers are violet, with red bracts.

A. S. STRIATUS. Leaves are variegated, creamy yellow on the margins.

Billbergia

Very often the aechmeas and the billbergias are confusing, for at different times the experts have named plants one or the other, and then changed them around. There is a difference in appearance, although their behaviour, and the cultivation necessary, are similar.

BILLBERGIA NUTANS. This is probably the commonest of all the bromeliads, being seen very often in all sorts of homes, and brought to shows for naming. It has the usual urn, but becomes more packed, while the silvery, narrow leaves have slight spines and form clusters rather than rosettes. These plants are often sadly neglected. They will be put under the staging for weeks, and as the seasons pass the pot becomes overcrowded, while the tops become choked by the urns that have borne flowers. If they are kept young, the beautiful flower arrives on a slender rose-tinted scape, dangling in smallish bunches of green petals edged with blue. They are easily grown in a cool greenhouse, or living-room, but will love the warmth and company of the normal warm greenhouse occupants.

B. PYRAMIDALIS (syn. *Bromelia pyramidalis*). Leaves are lengthy, about 2 in. across and banded white. Bracts are pink. Flowers are in upright spikes and are a mixture of red and purple with white, mealy sepals.

B. SAUNDERSII. Longish leaves form a deep narrow rosette, the surface spotted and mealy to give a white appearance. The flowers dangle in a mixture of crimson, yellow and blue.

B. WINDII. Very similar to *B. nutans*, being an offshoot of that and other species, but larger and more mealy.

Note. Other names in trade lists are: *B. distachia*; *B. hertzii*; *B. hoelscheriana*; *B. leopoldii*; *B. vittata*.

Cryptanthus

This is a smallish genus of bromeliads noted for flattish plants that form the usual rosette cup, with flowers sometimes submerged in the water. They need much the same conditions and treatment as aechmeas and billbergias.

CRYPTANTHUS BIVITTATUS. A remarkable plant with long lower leaves almost flat on the pot. They have spiny margins and are striped brown with pinkish green shadings. The white flowers are insignificant.

C. ZONATUS (syn. *Tillandsia zonata*, *T. zebrina*). The leaves suggest a starfish, but colourings are brown, cross-banded in differing directions with white and silvery green. They curl over the sides of the pot as though enveloping it unto itself. There are hybrids and varieties in cultivation.

Guzmania

Those I have seen of this genus have a dense cluster of smallish leaves around the rosette, from which grows the flower scape. They are more horizontal than are those of aechmeas. Most of the species are epiphytes, and should be grown the same way as aechmeas.

GUZMANIA 'Fantasia'. This is a variety, probably of *G. cardinalis*, that was on view at the Royal Horticultural Society's Chelsea Show in 1972. The leaves formed a spiral clump, and from the middle grew a tall strong spike of brilliant red flowers that were reflected in the red tones upon the leaves.

G. MONOSTACHYA. Has long narrow leaves forming the rosette. Flowers arrive on an upright spike, and are white with bracts that are a mixture of red, purple and green. There are varieties.

G. ZAHNII. Curving leaves up to 2 ft. long, forming the rosette. They are almost transparent and have stripes of bronze and red. Flowers are light yellow.

Neoregelia

These have been grown under various names, according to the species, including *Nidularium*, *Karatas* and *Billbergia*. There are several in cultivation, the chief attraction being the way the inner leaves turn to brilliant colours. Like nidulariums, their flowers are quite near the water in the rosette, and are sometimes submerged. They need similar cultivation to aechmeas.

NEOREGELIA CAROLINAE (syn. *N. meyendorfii*, *Bromelia carolinae*, *Nidularium meyendorfii*). Has a large rosette of leaves that are bright green above and lighter below. Flowers are violet, with red bracts to form a cauliflower. Inner leaves turn red at flowering time.

N. CONCENTRICA (syn. *Aregelia concentrica*, *Karatas acanthocrater*, *Nidularium acanthocrater*). Has a large rosette of dark green leaves that are grey-banded below. The bird's-nest middle has a surround of purple that is spotted brown.

N. MARMORATA Hybrid. Has leaves that are marked red and green. Lavender flowers appear in the rosette.

N. PINELIANA. Has narrow leaves, forming a neat rosette, that are marked and spotted red or brown. Flowers in the centre are violet.

Note. Other species or varieties in lists are: *N. ampullacea*; *N.* 'Fairypaint'; *N. fosteriana*; *N. spectabilis*; *N. farinosa*; *N. oligantha*; *N. tristis*; *N. zonata*; *N. sarmentosa*; *N. regelioides*.

Nidularium

These are more spready relations of aechmeas, looking more like cauliflowers. The blossoms of the species are sometimes heaped, but sometimes do not appear above the rim of the cup. The outer leaves are generally green, although they have darker undersides. The chief attraction is the broad inner circle of leaves that are of an intense glossy redness, as though varnished. The cup in the middle is definite, and water that accumulates should be left. Their needs are much as for aechmeas, i.e. cosy warmth, moisture and slight shade from sunshine.

NIDULARIUM FULGENS (syn. *N. pictum*, *Guzmania picta*, *Karatas fulgens*). A compact grower, with outer green leaves, each about a foot long and mottled, with prickly edges. Inner leaves are intense scarlet, through which peep, and also from the middle, violet flowers that have white tubes.

N. INNOCENTII. Outer leaves are stout and broad, dull green, with a central disc of intense red. Flowers form a white rosette in the middle.

N. RUTILANS. Vermilion flowers among the inner leaves and in the cup. Inner circle of leaves are handsomely red with an overtone of rose.

N. SCHEREMETIEWII. This is a large species, often with a spread of 18 in. Inner circle of leaves are bright red. Flowers are purple with a suggestion of white.

Note. There is a lot of confusion in the naming of many nidulariums. Some were at one time transferred to a genus named *Aregelia*, but have since been changed to *Neoregelia*. *N. meyendorffianum* could now be listed as *Aregelia carolinae*, *Neoregelia carolinae* or *Bromelia carolinae*, if not as above. Still another genetical name is *Canistrum*. Some were named *Karatas*, but this genus has been eliminated.

Puya

These, like the commercial pineapples, grow in soil. They are very spiny, greyish green, are evergreen, and presumably perish when the growing cycle is completed. I once had a greenhouse specimen that did not bear flowers for eight years. It was not altogether attractive and was too spiny for comfort. Should be grown like a succulent.

PUYA ALPESTRIS. Has blue flowers with orange anthers. Suitable for a cool greenhouse also.

P. VIOLACEA. Has leaves that are long, narrow and recurved. Bears violet-blue flowers on a very tall stem.

Note. Other genera similar to puyas are *Dyckia* and *Pitcairnia*.

Quesnelia

These are closely related to the aechmeas and billbergias, and mostly have long thinnish leaves with spines. If treated as advised for aechmeas they will thrive.

QUESNELIA LIBONIANA (syn. *Billbergia liboniana*). Grows about 1 ft. high and bears a spike of flowers of red and blue, during winter.

Q. ROSEO-MARGINATA (syn. *Billbergia roseo-marginata*). Grows about 18 in. high. Has leaves about 2 ft. long that are banded and mealy, with some spines. Flowers come in a spike of light blue with rosy bracts.

Tillandsia

There are a few hundred species of this genus distinct from other genera by certain botanical differences but, generally, for our purpose they are allied to the vriesias and guzmanias.

Most are epiphytes, and need no solid compost in the pots, the treatment for aechmeas being suitable.

TILLANDSIA DURATII (syn. *T. circinalis*, *T. wallisii*). Grows about a foot high. The leaves are heavily coated with silvery meal. They emanate from the middle after the manner of celery, forming thick rounded leaves at the base that gradually narrow to a tip curled over to form a circle. The spikes of flowers are a very bright blue colour.

T. LINDENII. Probably the most beautiful of the many kinds. The recurved leaves form a lovely olive-green rosette, from which the scapes grow and form intensely blue flowers amid rosy bracts. There are several varieties.

T. USNEOIDES (syn. *Dendropogon usneoides*). This is a peculiar member of the genus, having no roots; the usual way of growing it is on bark. Place the base of the plant against the bark, put a roll of moss over that and tie the lot with raffia. Strong, dangling tufts form, that become covered with a silvery crust, giving the plant the common names of old man's beard and Spanish moss. Unimportant flowers of a greenish hue arrive in late summer.

Note. Other tillandsias listed in trade catalogues are: *T. aeranthos*; *T. bergeri*; *T. bulbosa*; *T. butzii*; *T. caput-medusae*; *T. dianthoidea*; *T. erubescens*; *T. flabellata*; *T. nubis*; *T. pohliana*; *T. pulchella*; *T. stricta*; *T. tricolor.*

Vriesia

There are about 100 species under this name. Their form of living is akin to that of aechmeas. They have the usual rosette, the leaves being variously coloured. Flower scapes are midway between the stiff aechmeas and the nodding billbergias.

VRIESIA HIEROGLYPHICA (syn. *Massangea hieroglyphica*). This is a giant with long, succulent leaves about 4 in. across that are much variegated with dark green shadings and undersides purple. Flowers are chrome yellow and less spectacular than those of aechmeas.

V. SPLENDENS. Leaves are smaller than the above and are banded green with brown shadings. The flower spike is erect, bracts being bright red; the actual flowers are yellow.

Note. Other species are: *V. carinata*; *V. duvaliana*; *V. psittacina*; *V. tessellata*.

Further Note. Other genera not reviewed here include *Androlepis*, *Bromelia*, *Canistrum*, *Chevalieria*, *Catopsis*, *Disteganthus*, *Fascicularia*, *Gravisia*, *Greigia*, *Hohenbergia*, *Lidmannia*, *Ochagavia*, *Portea*, *Ronnbergia*, *Streptocalyx* and *Wittmackia*. No doubt in time the British Bromeliad Society will establish all the names of the general and species. My chief task here is to make the reader aware of the extent of an amazing tropical family that will thrive in the stove or warm greenhouse. Most of them do not mind which.

CHAPTER 11

Euphorbias (Euphorbiaceae)

THERE are some 5,000 species of the spurge family, ranging from troublesome weeds to a few special subjects for the warm greenhouse. Their general characteristics are a milky sap that is often poisonous, and rather insipid flowers that are more than offset by glorious bracts. Most popular among these are the poinsettias that arrive at Yuletide to enhance the evergreens. The majority perish. But we can, and do, grow the old types for many years, when we have the facilities. We also have in this family the crotons and a few other first-class genera, chief of which are mentioned here.

Acalypha

Acalyphas are not as refined as crotons, but will tolerate somewhat cooler conditions.

ACALYPHA HISPIDA (syn. *A. sanderi*). The red-hot cat's-tail, or chenille plant, has as its chief attraction bright dangling tassels, somewhat like refined love-lies-bleeding, which are about a foot long and arrive in summer. It needs the moisture of the stove, and normal compost and treatment.

A. WILKESIANA (syn. *A. tricolor*). The copper leaf, as it is commonly known, is grown for the metallic green leaves, marked and blotched crimson-red.

A. W. MARGINATA. Largish leaves have reddish brown middles, with carmine colouring towards the edges.

A. W. OBOVATA. Leaves are green with creamy edges deepening to crimson.

Codiaeum

The codiaeums, or crotons, are much variegated, evergreen shrubs, that are comparatively easy to grow, provided they can get consistent warmth and plenty of moisture, such as the stove provides. They grow fast in a peat compost, but the leaves are better coloured when John Innes compost-cum-peat is used. By potting-on and feeding, it is possible to have specimens 6 ft. tall and 6 ft. across, but what we normally see are single-stem representatives just a few feet tall, looking like refined metallic aucubas in multitudinous shades of colourings. One of the problems is keeping them moist, for if at all dry, red spider mite spoils the lower leaves, leaving bare stems. This effect is countered by continually renewing the root parts. When a plant is 3 ft. high, say, cut the top growth off about 15 in. below the tip, just underneath a leaf. Remove enough leaves to plant the lower end in a 6-in. pot and put it in the close propagator. It will soon form new roots and grow nicely. If doubtful, try the method described for poinsettias. The lower part will, if left, produce new side-growths and be on its way to becoming a bush instead of a single-stemmed specimen. If bushy plants are wanted, the procedure is to cut them down from time to time. The usual way to increase them is by selecting the best coloured shoots, and rooting them in the propagator.

CODIAEUM EVANSIANUM. Has wedge-shaped leaves that split into three lobes. Young leaves are pale green, heavily marked yellow, and deepen with age.

C. JOHANNIS. A large-leaved kind with deep yellow markings.

C. PICTURATUS. Leaves are of variable shapes up to 15 in. long. They have red midribs, but the general pattern is green blotched yellow, that turns red with age.

C. VARIEGATUM PICTUM. This name is given as an indeterminate parent of the hundreds of varieties and cultivars in cultivation. There are other names, covering some of the forms that have been around a considerable time, which are now considered to be species in their own right and are here included as such.

C. WARRENII. Leaves quite 2 ft. long and very broad, much mottled yellow and red.

C. WEISMANII. A close-growing type with narrow green leaves striped or streaked brilliant yellow.

C. WILLIAMSII. The strong leaves are saw-edged and are a fusion of intense green, yellow and red.

Euphorbia

EUPHORBIA FULGENS (syn. *E. jacquinaeflora*). This near relative of the poinsettia is a more attractive subject, with its many bunches of smaller, red-orange bracts, studded along the arching branches and shoots. It needs the same conditions as poinsettia but prefers the stove in the earlier stages.

E. MILII SPLENDENS (syn. *E. splendens*). A species from Madagascar, commonly known as the crown of thorns, making an evergreen, succulent shrub with spines, and bears scarlet flowers and bracts almost continuously. Height is about 4 ft. It loves sunshine and warmth. Compost should be open and porous. It grows steadily and needs repotting each spring, not necessarily into a larger pot.

E. PULCHERRIMA. The poinsettia, being Mexican, requires a good winter temperature, and once the bracts lose their lustre and the leaves fall in the New Year, they should be gradually dried off at the roots and kept unwatered in the warm greenhouse. In April, the old dead stems are cut down close to the base and some water is given to the roots. When new shoots start, the plants are taken out of their pots, the soil carefully shaken off the roots, and any straggly roots shortened back. They are then repotted into the usual John Innes mixture-cum-peat, but should never be put into larger sizes than necessary. More water is given as they grow, and they are repotted into larger sizes as needed. The plants like shading from hot sunshine, and appreciate syringing with tepid water. They are placed outdoors from July until September, a sheltered frame that is shaded during the heat of the day being good. They are brought indoors again in the autumn.

Propagation is by slips pulled off the old plants in spring, each with a heel attached. These root better than straight cuttings, owing to the milky sap being a deterrent. Only the loose bark is trimmed off. They root within three weeks if placed in sandy peat in the close propagator.

A young plant of the old species, as apart from modern dwarf varieties, is inclined to produce one tall stem that will duly bear a large bract. Some gardeners cut halfway through that stem about 15 in. below the top. This callouses over within a fortnight. Then the rest of the stem is severed. The top part is then planted in a 6-in. pot and placed in warmth and kept moist until it roots properly. The old plant throws up side-growths and will bear several bracts later in the season. This operation is done in early September, when outdoor plants are brought back into the warm house.

CHAPTER 12

Ferns

BOTH ferns and palms need much study in order to know all about them. Here we are only concerned with their usefulness in the stove and warm greenhouse as subsidiary companions to the more general collection of ornamental plants.

At one time ferneries were shady pits more favoured by gardeners than their owners, who often merely acquiesced to their existence.

The jargon used when talking about ferns particularly was, and is, different to that employed for other plants. For example, the leaf of a fern is invariably referred to as a 'frond'. Its seed is a 'spore'. It bears no flowers, but the spore is bred and born in a capsule on the frond and is one of a 'sorus'. The way in which these are grouped is often denoted in learned books to distinguish the various species, of which there are at least 10,000 different ones.

Probably the easiest way to recognise this form of procreation is by studying the back of a leaf of the common hart's-tongue (*Phyllitis scolopendrium*), which grows freely under the trees in some parts of Britain. Whereas most ferns have very elaborate fronds much serrated and patterned, this one produces a solid blade containing a stiff, raised midrib. At right angles to the midrib are to be found the sori (plural for sorus), running parallel to one another. These are not so pronounced upon other ferns, but all have them in varying ways, and if we want to propagate plants from spores, they can be carefully harvested, just as are seeds from normal plants.

Propagation

The usual procedure is to keep a watch on the plants, and when the sorus turns brown, the frond, or fronds, bearing it is cut off and placed in a bag that is hung in a dryish, warm, shaded place, where the ripening will be completed and the spores shed by the frond. They are like fine powder, and can be kept in that state for a little while, but the earlier sown, the more likely are the chances of successful germination.

The spore-pan is prepared in a way much like that described for raising gloxinias. Best receptacles are of clay, well prepared with drainage crocks, and filled with Irish moss peat mixed with silver sand. The peat should be well soaked and then partially dried again before the sand is mixed with it. Press the contents down firmly with a tamp, or the bottom of a flat pot. Then scatter the spores on the surface, but do not cover with more compost. Rather should the pan, or pot, be placed in another much shallower pan, or bowl, with about ½ in. water in it, so that the water just reaches the level of the bottom of the compost in the spore-pan, so keeping the spores moist by capillary attraction (see Fig. 7).

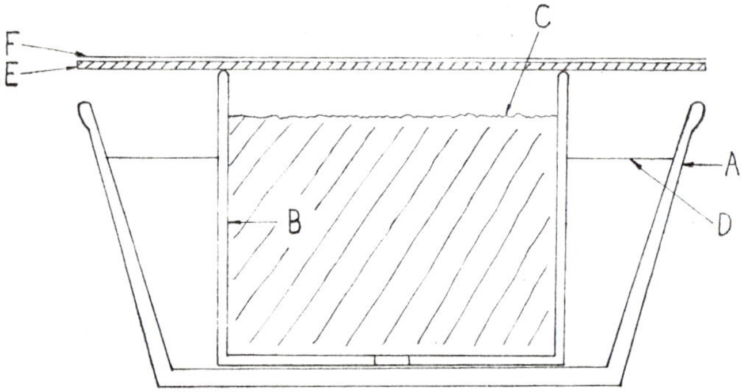

Fig. 7. Propagation of Ferns by spores.

A—Outer pan.
B—Spore-pan.
C—Level of surface upon which spores are sown.
D—Level of water maintained in outer pan.
E—Pane of glass over spore-pan.
F—Covering of paper over glass.

Put glass over the pan and stand the whole in the warmish propagator, covering the outside with newspaper, for preference, to keep out strong light. If these spores are sown in late summer, they will germinate during early autumn, late autumn, or even winter, if in normal propagator heat. Keep them moist by adding water to the saucer when necessary, and keep them shaded.

In time, there will be perceptive signs of growth. Directly you can see them sufficiently, which could be early in February, they are lifted out of the propagator, if this is wanted for other things, and placed in the stove in a shady place. The main point is not to give any overhead watering; it washes the spores away.

The resulting tiny plants will be so small that they may be difficult to separate. Sometimes individuals can be prised out with a pointed label or tweezers without damaging the roots, but more often they will come out in clumps of three or four. It is best to transplant them into one of the modern plastic seed trays, spacing them about 2 in. apart. Whether single plantlets or clumps, do not try to separate the latter, but plant together, as one would lobelias. They are inserted in the compost very lightly.

For this the mixture should be a little lighter than for the normal pottings, or repottings. If the usual John Innes compost is used, and the amount doubled by adding Irish moss peat, the texture will be more or less right. When using Irish peat, or any of the no-soil composts, always soak well first.

After the first transplanting, do not water overhead. Instead, place the trays in bowls or saucers containing just enough water to seep upwards to the plantlets through the drainage holes. Never have it deep enough for the roots to be immersed. Place the pans in the propagator, or a close atmosphere in the house, and shade from sunshine.

Once the small plants appear to have recovered, they can be moved to the greenhouse and put in a shady place where there is plenty of humidity. They are never sprayed. If the roots become dry, give water by steeping in a bowl for a few minutes (see also Potting-on).

Potting

When large enough, each small clump, be it of one or more plants, should be lifted intact and put into a 4-in. pot. The mixture for this can be the ordinary John Innes compost-cum-peat, this time adding one-third peat to the bulk, not forgetting to soak the former first.

When the small pots become filled with roots, the plants are put into larger ones. Drainage should be exceptionally good, using plenty of crocks over the holes in the bottom of pots, and rough peat over the crocks. Disturb the roots as little as possible when transferring from small pots to larger ones. Never use pots that are much larger than the previous ones. Push the compost in gently, and firm with the fingers much more moderately than is usual for other warm greenhouse occupants.

Composts

If anyone cares, a compost other than the general one just described can be used. For instance, leafmould is often recommended for ferns. It is natural and probably better than peat, if the other ingredients are mainly loam and sand. One recipe is equal parts loam, leafmould, peat and sand. Another is 2 parts peat, 1 part loam, $\frac{1}{2}$ part sand, and a little charcoal. Another is 2 parts loam, 2 parts leafmould, 1 part sand and 1 part dried manure with 1 tablespoonful of bonemeal and 2 tablespoonfuls of small charcoal to each peck of the mixture. In reality, ferns are not as fussy about the ingredients as we would pretend. But they should have some food, plenty of drainage, and should never be grown in sour soil. This happens when they are given too much room at transplanting time.

Watering

Established plants should never lack water. They need it in abundance throughout summer. They need less in winter, and have much in common with evergreen shrubs in that respect. They can suffer from too much wet or dryness. As a general rule, never use a syringe on them; they are too delicate. The humidity should be in the air. They abound in shady dampness, so often sheltered from the rains.

Potting-on

Established plants are potted in early spring. They resent root disturbance, but are more tolerant then than at any other time. There are two main groups: some have crowns and remain static; others have creeping rhizomes. If of the latter sort, the pans need to be larger and larger each spring. If of the crown type, the pots need only be slightly larger.

Should it be difficult to raise new stocks from spores, as is frequently the case, the crown types can be divided. The rhizomes of the other can also be cut into pieces. The essential thing is only to pot up pieces of rhizomes, or crowns with rootlets attached. Having done that, place them in a propagator until they are growing freely once more.

Feeding

Old plants, once their roots have filled the pots, need feeding. We used to soak cow manure and sheep droppings in tubs of water in days gone by. Now we get bottles of Liquinure, or Maxicrop, or some such carefully blended fertiliser. I think it pays to ring the changes with these. Plants, like us, need a change of diet from time to time.

Pests

Unfortunately, ferns are, if anything, more prone than other plants to the usual warm greenhouse pests—woodlice, ants, mealy bug, scale and red spider. However, if the moisture is sufficient, the last-named cannot live in it. If the rest are kept out of the greenhouse by careful screening of every new plant brought in, it is unlikely that there will be an attack of mealy bug or scale. Take care to distinguish between spore cases and pests. The former are sometimes mistaken for the latter.

Among the antidotes for other plants I have advocated is, because of its efficiency and non-persistence, malathion in aerosol form. Alas, this is likely to damage the finer foliage of all ferns, and some of the soft-leaved plants. Therefore, either use an insecticide that will control all plants less effectively, or take care to use malathion only upon those plants it does not damage. In any case, if used in smoke form, rather than as a local aerosol spray (which is the best and most economical way to do it), all the ferns must be removed from the house and not returned for at least 12 hours.

My preference is always to have a malathion aerosol on hand, to deal with any local outbreaks of any insect upon the general run of other plants, but for ferns and tender subjects to use either a nicotine wash, or a fumigator. Alternatively, if the house is adjacent to living rooms, such as a lean-to with the entrance to the lounge, say, a non-poisonous brand, such as Bentley's Premier Plant Spray, is safer.

I regard nicotine as being too dangerous. But in all cases, great care is necessary. No matter whether an insecticide is regarded as non-poisonous or not, all are toxic in one way or another, or they would not do the job. Always heed the advice upon the labels.

Eelworm. One of the most serious of fern troubles is eelworm. Its presence is seen by the fronds (not the stems) of maidenhair and other genera being disfigured by definite and clear cut areas of brown or black. All such leaves should be taken off, close to the soil surface, and burned. These blotches are the breeding ground and if they are not removed, the eelworms eventually go down into the compost in the pots, and so spread their activities. Very badly affected plants, with the pots and compost, should be completely destroyed.

They are very difficult to eradicate. One way is to prepare a bath of liquid nicotine soap, 1 pt. of the bought preparation making 4 gal. of dip. Immerse the plant and pot on its side in this for several minutes, and then stand in a shady place to dry out.

Main precautions are never to use old pots unless they have been properly sterilised. This can be done by steam, or by soaking in an effective disinfectant. I have found that a strong dip of boiling water, to which is added a tablespoonful of Dettol for every 2 gal., has rid the pots of pests and fungus. Incidentally, fungal troubles are less severe in stove houses than those with lower temperatures, or outdoors.

Brake Ferns

The best known wild fern is bracken, which grows in many parts of Britain in overwhelming quantities. It has scores of relatives under many difficult names. But the type is generally referred to as brake fern, the largest genus being *Pteris*, a word of ancient origin meaning a feather, therefore, we can also refer to them as feathered ferns.

As a matter of interest, fern experts are referred to as pteridologists, and their science as pteridology. They tell us that of the 10,000 species living upon this planet, only 250 belong to the genus *Pteris*. However, there are about 170 genera, and 7,250 species within the family *Polypodiaceae* to which they belong. The names are changed around from time to time. Seeing that, according to them, the maidenhair is a relative of the same family, I can only appreciate the difficulties they become involved in. As an example, our woodland bracken, that had, or has, common names like adder-spit, brake and eagle fern, was once *Pteris aquilina*, but is now found in dictionaries as *Pteridium aquilinum*. *Pteris palmata*, that was, is now *Doryopteris pedata palmata*, unless it has been returned to *Pteris* again (it was originally changed from *Doryopteris* to *Pteris*). We must let the experts wrangle, and try to be sure about the plant that we are discussing.

Adiantum

The commonest greenhouse representative of this large genus needs no description, for everyone knows maidenhair fern when seen in floral tributes of every description, together with smilax and *Asparagus plumosus*. Actually, these last two are not nearly related to ferns, but are members of the lily family under which heading they are dealt with.

When I was first put to watering the plants I was fascinated to see that water sprayed over the maidenhair plants formed large drops and fell off, leaving the fronds completely dry. The naming authority noticed that too, for the name adiantum means 'dry'. The plants love moisture around and among the roots, but never should it be syringed upon the foliage.

ADIANTUM CUNEATUM. Everyone recognises this, as the maidenhair fern, but can never be sure when it is the true species. There are very many varieties and cultivars. The original has fronds about a foot long, with branches spreading to about 8 in. Clear stems are about 8 in. long, very wiry, and are referred to by experts as stipes.

The usual habit is to form a crown, from which grow numerous fronds practically throughout the year. If allowed, those bearing sori will mature and produce spores very freely. Quite often these yield self-sown plants on a damp floor or where germination can take place. The tufted crowns can also be divided at potting time if necessary.

The John Innes compost-cum-peat makes a good rooting material. This species will be happy in the warm greenhouse or the stove, as long as the air is humid, and an occasional feed with one of the proprietary bottled fertilisers will help when fronds are being produced in large quantities.

A. C. GRANDICEPS. I mention this one because it has crested fronds that droop, and plants are often used in hanging baskets.

A. C. MUNDULUM. This is a very dwarf variety with stems not more than 3 in. high. It is a real miniature, looking much like a fuzzy green mat, and looks good in a pan, or as an edging to staging, or in hanging baskets.

Note. These are just two examples of sorts that are good for purely ornamental purposes, rather than for cutting the fronds, as is the case with the true species, or its type.

A. FEEI. Comes from Mexico and has strong upright stems a foot long or more, with large, spreading fronds and beautifully shaped leaves. It needs the stove. Not as easy to grow as *A. cuneatum*, by any means, but an aristocrat of the genus.

A. MACROPHYLLUM. Has black stems up to a foot long, with fronds about the same length and 6 in. across, and needs stove conditions. It looks less like a maidenhair than most. The oval leaflets (divisions), neatly patterned up the stem, vary from 3–4 in. long and are 2 in. broad. There are varieties of this, with white stripes in the leaves.

A. CAUDATUM. The trailing maidenhair is also a good fern to grow in the ornamental stove, if you can get pieces, or spores, of it. It is more accommodating than the last two, and will flourish in either the stove or the warm greenhouse. The stems, or stipes, are short and wiry. The fronds droop over until the tips touch the soil, when they form roots below and plantlets above.

A. TENERUM. A grand species with very long fronds that are equally broad.

A. T. FARLEYENSE. Considered by many gardeners to be the grandest of all maidenhair varieties.

Note. All the above need the stove if they are to grow well. The list is merely a cross-section of the species and varieties.

Aneimia

These are novelty ferns, the chief attraction being the arrangement of the panicles, which bear the spores in sori looking more like blossoms than is the case with any other ferns. There are stove species, and also those that will thrive in warm greenhouses. They grow in the usual John Innes compost-cum-peat quite well, and need small pots, for they seldom grow to considerable size. Leafmould in place of loam is also recommended by some authorities. These are often grown in living rooms in Wardian Cases, but we are more concerned here with them blending in with other subjects of the greenhouses. They love light, and make splendid spacefillers between larger plants. In other respects the general treatment for ferns suits them. Do not wet the fronds.

ANEIMIA ADIANTIFOLIA. Resembles the maidenhair to some degree. Has long stalks, headed by spreading leaflets.

A. DREGEANA. The ash-leaf fern has long stalks (stipes), with fronds about 9 in. long with leaflets (divisions) slightly resembling the ash.

A. PHYLLITIDIS. (syn. *Anemidictyon phyllitidis*). The flower fern has long stipes surmounted by fronds, that are often marked in various shades of green. Has a pronounced panicle, that attracts much attention. This species will thrive in a warm greenhouse.

Asplenium

Here we have a genus comprising over 500 species of very differing characters and forms. They are, for the most part, tougher than most other ferns, and less subject to pests. They vary in size, the larger ones being the easiest to grow. Some are wildings in our own country, but many definitely need the warmth and moisture of the stove.

ASPLENIUM BULBIFERUM. The mother spleenwort has finely pinnate fronds about 18 in. long, upon which bulbils form and then develop into tiny plants, when they then drop off and root in the soil. This species is very popular, and much grown in parlours and lounges, such conditions being to their liking. The warm greenhouse is much better than the stove. They can be propagated by taking off the bulbils before they want to drop, planting them in pans of sphagnum and peat mixture. When nicely sizeable, they can then be put separately into small pots of John Innes compost-cum-peat.

A. LONGISSIMUM. Has a mere stipe (stalk) about 6 in. long and evergreen fronds up to 8 ft., that have a fishbone series of leaflets, well covered with sori. The tips form roots when they touch the soil. This one needs a lot of room, and is not recommended except for a large stove.

A. NIDUS. The bird's-nest fern makes an unusually attractive plant in a pot while young. It has entire, instead of laddered, leaves, after the manner of hart's-tongue, but they are much broader across the middle and more wavy. They grow from a central stub to form a closely packed circle at the base, each eventually attaining a height of 3–4 ft. on old plants. The best position is the stove, where there is plenty of light but not direct sunshine. The plants grow better if moved to the cooler greenhouse during summer, and can stand outdoors for a few weeks against the north side of a greenhouse, where the sun only reaches them very early in the morning and again in late evening.

The most suitable rooting material for these is a blend of sphagnum and Irish moss peat, well wetted and then dried. These ferns should never be potted very firmly. Best means of reproduction is spores.

A. N. MUSAEFOLIUM. The fronds of this one can attain 6 ft. and be 1 ft. across the middle.

A. RADICANS. Another type of spleenwort. It has a caudex (main stem or bole, so becoming a standard) about 1 ft. long, and sprouting from its top are fronds some 4 ft. long with stipes (clear stalks) about the same length as the caudex. The leaflets of the fronds are up to 2 ft. across at the base, and taper gracefully to the tip, forming a beautiful wedge.

Lygodium

These are closely related to the aneimias, but behave quite differently, being the only genus, to my knowledge, that really climbs. Placed against greenhouse pillars and posts, or given string to twine round, they will attain considerable heights. Not only do they wind themselves round the supports, but also their own stems, so forming a mat of fronds of various forms of leafage. They grow better in peaty borders than in pots.

LYGODIUM PALMATUM. Grows like a vine and has fronds that slightly resemble it, rather than the fern image. The spores are produced on separate growths during autumn. Needs a warm greenhouse.

L. SCANDENS. Produces long-thread-like growths clothed with attractive foliage that is a delightful pale green. It will produce long, thread-like shoots that are most useful in cut flower arrangements. Thrives in the warm greenhouse.

Other excellent species are *L. japonicum*, *L. polymorphum*, *L. reticulatum* and *L. volubile*. At one time these genera were under other names that included *Hydroglossum*, *Lygodictyon* and *Ophioglossum*.

Nephrolepis

The species of this genus, generally referred to as ladder ferns, are definite contrasts to the maidenhairs, and look more like the general image of the family. Purely for decorating the ornamental warm greenhouse, they are preferable to the more popular maidenhair. The fronds are straightforwardly feather-shaped, with leaflets (divisions) appearing evenly throughout the length, which is often very considerable. They are not difficult to grow, needing much the same treatment as the maidenhair.

NEPHROLEPIS ACUMINATA (syn. *N. davallioides*). Produces a bouquet of fronds, leaves starting about a foot above soil level on pliable stems, the outer ones drooping gracefully to form a handsome cluster. They extend to 2 ft. and sometimes 3 ft.

N. A. FURCANS. This is superior to its species, possessing more character, the fronds dividing at the extremities, remindful of earwig's tails. They are crested and over 3 ft. long, arching gracefully.

N. EXALTATA. This species has long narrow fronds up to 2 ft. high that grow thickly together, forming a mass of fine greenery. It will flourish under all sorts of conditions.

Pityrogramma

The chief characteristic about this genus is that the fronds are dusted with golden or silver powder. The different species vary in the formation of the fronds, some being filmy, and others whose divisions are practically solid. They appreciate less moisture than most ferns, but too much sunshine spoils the mealy hues. They should not be syringed, but kept supplied with water at the roots and they like the stove conditions.

PITYROGRAMMA CHRYSOPHYLLA. (syn. *Gymnogramma martensiana*). The golden fern grows about 1 ft. high and is powdered bright yellow.

P. PERUVIANA ARGYROPHYLLA. Has reddish stalks and petals, and leaves dusted with silvery powder.

P. PULCHELLA WETTENHALLIANA. This lovely, finely silvered fern has fronds with tassles. There is some confusion between the species and the variety. This and the preceding were, and are, alternatively named *Gymnogramma*.

CHEILANTHES MYRIOPHYLLA ELEGANS. Related to *Pityrogramma*, but has rather small fronds that are very finely divided. They are hairy and of a lovely green shade. Makes a beautiful miniature specimen in the stove.

Platycerium

These comprise the most startling and unusual form of the fern family, differing from all others in several respects. They are beautiful after their own fashion, and wherever seen arouse intense interest. To some extent they resemble those mounted antlers hung upon the walls of sporting gentry, and are either grown in a basket, or fastened to a piece of cork or wood.

They are epiphytes, which means that in the tropical jungles they hang upon trees, without drawing sustenance from them. There are two forms of fronds. The lower ones are sterile and clasp the host, spreading out at the top so that leaves and falling rubbish from the tree collect between their outer edge and the bole, so making a natural compost into which the roots develop. They also depend upon the humidity and warmth of the atmosphere. The sterile fronds are a lovely green when young, but eventually turn pinkish red and become more like coarse paper.

The fertile fronds are green, overlaid with a fine greyish down on the under surface, and splay out in the pattern of a deer's or elk's horn. The sori are borne irregularly upon the undersides of the fronds near the forked tips.

Some growers try to emulate the natural conditions by fixing them to pieces of cork bark or wood, and wrapping peat and sphagnum around the base. They will grow quite well in small baskets, and can easily be fixed inside them, when filled with a mixture of peat and sphagnum.

The easiest way to propagate them is to wait until the roots form a spongy ball at the base, and show adventitious buds. These are hacked away with some of the root and, when placed in a pan or pot of sphagnum and peat, will slowly but surely develop. They are slow to grow, and stay in good shape for many years. It is also possible to raise from spores in the usual manner.

They cannot be watered, or syringed except with a very fine nozzle. If the contraption in which they are growing, be it a basket, or a shelf of wood, or bark, is hung where the plant gets light without bright sunshine, such as on the north side of a wall, or shelf, it will survive. When it begins to look a little dried out, you simply take it down, dip in a bath of water long enough for it to soak through, then allow to drain a little while and put it back.

PLATYCERIUM BIFURCATUM. The elk's-horn fern, or stag's-horn fern, is the best known species. It will thrive in a warm or cool greenhouse. The forked fronds of simple composition attain a length of 2 ft. or more, and spread in all directions, in the style of a bouquet, without the trailer.

P. GRANDE. This is a 'grander' version of the preceding species and more difficult to grow, needing plenty of warmth and humidity. It does not produce the root buds, and has to be propagated from spores. The nodding, fertile fronds are in pairs and often 5 ft. long.

Other species are: *P. coronarium*; *P. hillii*; *P. stemmaria*; *P. veitchii*; *P. wallichii*; *P. willinckii.*

Pteris

PTERIS CRETICA. The Cretan brake makes a good house-plant, and loves the warm greenhouse, rather than the stove. Its fronds are about a foot long, and about half as wide. They stand up well on wiry stipes, and are yellowish brown in colour. Some of the varieties are quite interesting.

P. C. ALBOLINEATA. Has a white line through each leaf division of the fronds.

P. C. MAYI. May's brake also has the white line, as well as tassled tips at the end of the leaflets.

P. C. WIMSETTII. The Wimsett brake has the usual fern fronds up to 2 ft. in length, the chief difference being the chestnut red stripes on the leaflets.

P. ENSIFORMIS. The sword brake needs a stove house. This is a little taller and more pointed than the preceding. There are several variegated varieties, some with white lined, fuzzy leaflets.

P. MULTIFIDA. The spider fern is similar to *P. cretica*, but the leaflets are narrower, and many of the varieties are crested or plumed. Probably the best known is Rochford's Variety. These make good house-plants and only need warm greenhouse temperatures.

* * *

POLYPODIUM AUREUM (syn. *Phlebodium aureum*). The hare's-foot fern is noted for the mustard shaded spore cases upon the backs of the grey-green fronds, that are up to 5 ft. long. Its roots spread, and a pan is preferable to a pot. It needs the warmth and moisture of the stove.

P. A. MAYI. The fronds of this are silvery green and most attractive.

Tree Ferns

The word 'caudex' was mentioned earlier to describe an asplenium and a blechnum with a bole or main stem. There are, in fact, many genera of ferns that have such stems, the most notable being *Dicksonia antarctica*, with a thick caudex (trunk), that will attain a height of 30 ft. At the top is the cluster of fine, arching fronds. It has to be grown in a cool greenhouse which, obviously, must be large and high.

I have grown it outdoors in one garden where, according to advice, we hosed the thick caudex every morning during summer fine weather, and grew it in a large tub that stood where it was shaded from direct sunshine.

Anyone wishing to have a grand tree fern, and who has a warm greenhouse of considerable size, can try the Mexican version, named *Cibotium schiedei*. Its caudex (trunk) will eventually attain 15 ft. if kept for several years, and it will have fronds up to 5 ft. long. They curve in a beautiful, graceful way, and its fellow species can be raised from spores which are freely produced, without being overwhelming, for several years.

Other genera containing most imposing tree ferns are *Alsophila*, *Brainea*, *Cyathea* and *Hemitelia*. Some need stove conditions, and others will survive in a greenhouse. Most will eventually outgrow their positions.

BLECHNUM GIBBUM (syn. *Lomaria gibba*) is a miniature tree somewhat like *Asplenium radicans*. The caudex will possibly grow to 2 ft. with age. Fronds are elegant and lace-like, and the effect is a lovely small tree fern.

CHAPTER 13

Figs (Moraceae)

ONE or two species of the India-rubber plant family give brightness to a general collection of stove or warm greenhouse plants, but they should never be overdone, for there are so many less common and more exciting sorts of plants available. Taken as a whole they make better house-plants, being less exacting in their needs than many other subjects that are otherwise grown in living rooms. The chief point here is to explain that the selection is considerable, if anyone wishes to get away from the commonest forms.

The main needs in the greenhouse are simple. They thrive in John Innes compost-cum-peat. Pots should be small. They need not be repotted more than once every two years or so. Growing plants need a little liquid fertiliser occasionally. Position should be where they get light but not direct sunshine. They like to be sprayed, or have the leaves sponged. Watering correctly is important; they only need water in moderation at all times, even in summer. During winter they should be allowed to become almost dry before any more water is given.

In time they grow too tall. Air-layering (marcottage), can be resorted to. Make a deep embrasure in the stem, about 15 in. from the top of the plant, wrap it round with wet moss and secure. Roots will form in the moss, and then this can be cut from the lower part and planted in another pot. Points to bear in mind are that the moss should be kept moist, and there should be left a small gap at the top to pour in water when necessary. This can be conserved to a large extent by covering with polythene or some such waterproof material (see Fig. 8).

FICUS BENGHALENSIS. This is the famed banyan tree that attains a great height, and spreads by the aerial roots growing down to the soil and rooting. Small pot plants are quite attractive for a few years.

F. BENJAMINA. The 'indoor birch' has small, pointed leaves, while the main stem is weaker, causing a weeping habit. Its foliage is glossy and attractive.

F. DELTOIDEA. (syn. *F. diversifolia*). The mistletoe fig has small, almost round, leathery leaves and a thin stem, and is rather short and spreading. It is referred to as the mistletoe fig because it bears berries somewhat like that plant.

F. ELASTICA DECORA. The India-rubber plant is known to

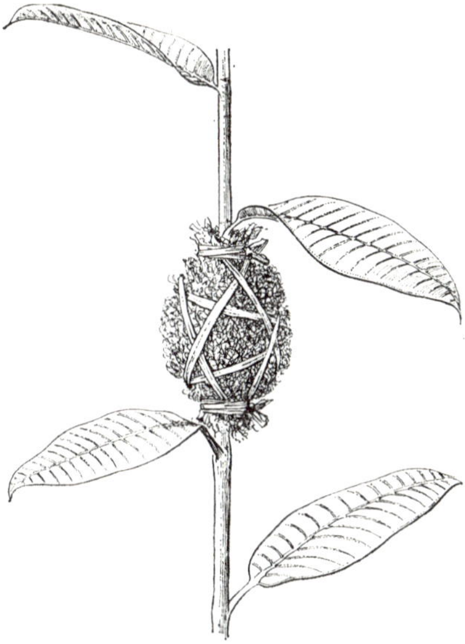

Fig. 8. Marcottage.

everyone. Light, without bright sunshine, causes it to have more deeply reddish growing tips than if kept in dull shade.

F. E. DOESCHERI. Is erect and has a thick stem, while leaves are elongated and rather tiered; they have a wide cream edge, and the whole leaf may be tinged pink.

F. LYRATA. The fiddle-leaf fig; the leaves of this species are much more interesting than the better known India-rubber plant. They are more waved, and shiny, and grow 18 in. long.

Many others within this range will be found in specialists' lists.

CHAPTER 14

The Gesneriads (Gesneriaceae)

IN my opinion, the family comprise the most beautiful of plants, if colour and shapeliness together are the criterion. Most have lovely velvety foliage that is alluringly exquisite, rather than grandiose.

Contained in the family are African violets, or saintpaulias. Then we have the delightful achimenes. In addition to that are genera or species that trail or climb, like columneas.

Coming from Brazil, as the majority do, the gesneriads love light, moisture, drainage and just a little shading, as though growing under trees, and fit perfectly into the normal stove or warm greenhouse conditions.

For compost the usual John Innes compost-cum-peat suits them very well. They like to be potted so that roots are fairly, but not too firm, while drainage must be good. They all love frequent doses of weak liquid manure when they are growing well, especially when forming new flower buds.

Propagation can be carried out by placing the leaves on compost, as described for begonias, or they can be inserted upright with the lower parts buried in a damp, sandy peat mixture and kept moist and warm (see Fig. 9). Small tubers, rhizomes, or roots will develop around the base, which can eventually be grown on for next season.

However, practically all are easily raised from seed in the way described in the first chapter.

Gesnerias as such exist under other names. Some are rechsteinerias. Others used to be naegelias but are now smithianthas, the consequence being that names vary according to the degree of knowledge or the whims of the catalogue compilers. Here the name given is the most recent, so far as is known. All the species are tuberous and respond to the treatment recommended for gloxinias. The main difference is that seed is generally sown in March instead of earlier. However, once there are permanent tubers, these need the dry winter rest. They can be started again from February to April.

There are a number of other genera with confusing names. It

Fig. 9. Gloxinia leaf with complete stalk inserted in 5-inch pan of compost (several leaves can be inserted in the same receptacle).

seems that botanists find differences in formation, and a few are changed from genus to genus according to opinion. Among them are *Isoloma*, *Tydaea*, and even *Achimenes*.

Achimenes

These beautiful flowers originally came from Jamaica and central America, and the hybridists soon produced many cultivars, some of which have been in cultivation for a century or even more. One such is 'Ambroise Verschaffelt', with white flowers that have a purple marking in the centre. Other newer and typical sorts are: 'Charm', pink; 'Mme Geheume', reddish-violet; 'Paul Arnold', light purple; 'Shirley Fireglow', red-orange; and 'Vivid', magenta-pink with yellow throat.

The general form of cultivation is to plant tubers in containers filled with the usual compost mixture needed for gesneriads. Sometimes the flowering receptacles are hanging baskets instead of pots. Very often the tubers are planted in largish boxes and transplanted when some growth is made. The tubers are spaced about $1\frac{1}{2}$ in. apart, and covered with just over an inch of soil. This is done from February to April, according to whether wanted in summer or autumn. They are kept moderately moist at first, in the warm greenhouse rather than the stove, and as they develop, more water is given, and if there are straggling growths they are nipped off. The outcome is a lovely mound of flowers up to 2 ft. high in colours according to their kind. Eventually, they tire and water is gradually withheld, until the wintering period when they must be kept in the receptacles, quite dry and warm, until repotting time. They are the easiest of the gesneriads to cultivate. Increase is by seed as for gloxinias, or by leaves and tubers.

ACHIMENES LONGIFLORA. Grows about 1 ft. tall and bears blue flowers during late summer. There are white and pink variations.

A. ROSEA. This species attains 1½ ft. and has scarlet flowers.

Aeschynanthus

This is an attractive genus of rather twining plants which like the stove house conditions, with a moist atmosphere and semi-shade. They should have very good drainage in the compost.

AESCHYNANTHUS GRANDIFLORUS. This is a plant that bears clusters of crimson and orange flowers looking like boat-shaped tubes at the extremities of long and straggly branches. Leaves are quite attractive and much more pointed than is usual in the family. Its natural way of life is akin to honeysuckle and ivy. If the shoots are pinched back while young and then pinched back again later, the resulting growths will be more bunched and give a lovely floral display. They need to be cut to the base immediately after flowering, which gives the roots a small resting period. Some of the new shoots in the spring can be rooted in the propagator to make new plants. At the same time old plants are repotted or planted in suspended baskets. These are lined with moss, and a light mixture is used, the plant being planted in the middle. As the shoots spread, so some are pegged down and, by judiciously pinching-back until flowers are wanted, a good shape is obtained. But always remember that they bear flowers at the tips.

A. SPECIOSUS (syn *A. aucklandii*, *A. splendens*). This species does not grow as straggly as the above, but does need much the same treatment, and mature plants can be kept in pots or wire baskets. The flower season is autumn and colour is bright orange.

Columnea

There are several first-class species in the genus, the general habit of which is to twine and trail, and if grown in pots and pinched back in the early stages they will, in due course, spill over to give wonderful bridal bouquet effects. They need identical treatment to aeschynanthus, but are easier to grow and can tolerate the temperatures of the warm greenhouse. One or two of the main stems should be cut right back each year, so that new ones are produced, with flowers nearer to the crown of the plant. The old stems unfold flowers further and further down, near to the tips.

COLUMNEA GLORIOSA. Makes a trailer down to 2 ft. long. Flowers are long and tubular, bright red with yellow throats. They arrive in early winter. There is a variety with purple-tinted foliage.

C. MICROPHYLLA. Has brilliant scarlet flowers, more trumpet-shaped than the above, and is a spreadier plant. It starts to flower in autumn, lasting for weeks. Mature plants can be developed into most impressive specimen trailers.

C. ROTUNDIFOLIA. The habit of this is little different, but leaves are smaller and flowers are larger.

There are several other species and varieties that appeal to collectors.

Episcia

These are much the same as columneas, and respond to the same conditions and treatment. The chief difference is that they have a resting period after flowering. They are more tolerant of a drier atmosphere.

EPISCIA DIANTHIFLORA. This grows into a dense hanging carpet that hangs down around a basket to a depth of 3 ft. This and the following are typical examples of several other species and varieties. Has whitish flowers with purple tinges.

E. LILACINA (syn. *E. chontalensis*). Has hairy leaves up to 4 in. long. Flowers arrive in axils of leaves and colour is mainly lilac. There are differences in shades of flowers and leaves.

Gloxinia,
see Sinningia

Saintpaulia
These revel in humidity. A good place for them indoors is a ledge just above the kitchen sink. Unlike the gloxinias, they never seem to go to rest, but will continue to flower over very

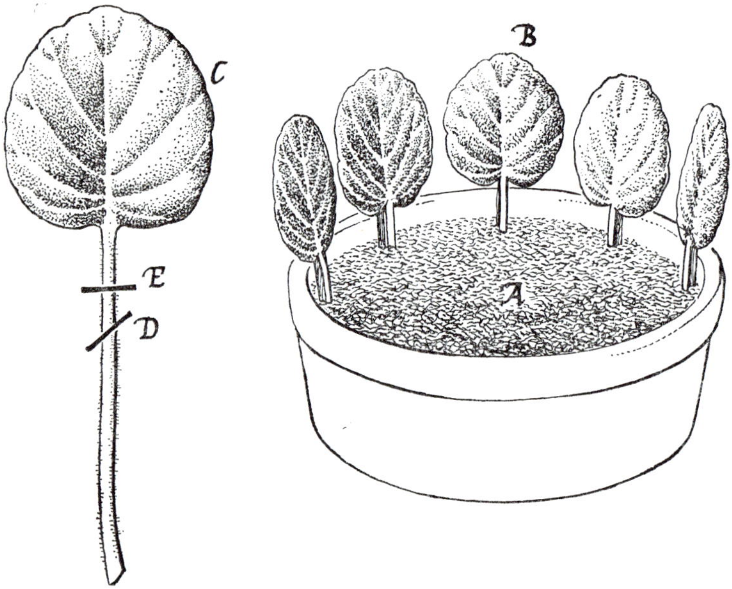

Fig. 10. Saintpaulia propagation by leaf.
A—Sand and peat compost.
B—Leaves with shortened stems inserted only $\frac{1}{2}$ inch deep and held erect by a toothpick or a matchstick.
C—Leaf with a long stem.
D—Where to cut with a sharp knife or razor blade.
E—Insert in compost to this depth.

long periods. They can be controlled to some extent by watering, or propagation. If kept moderately dry in summer, they produce winter flowers. If seed is sown in March, they will also be ready for winter.

One way to propagate named varieties is to insert shortened ends of the leaf stalks into compost, supporting the leaves with a matchstick or toothpick (see Fig. 10). In time rootlets will form around the bottom. After a few weeks, the tiny plants are lifted, carefully sorted out, and potted up singly in the John Innes compost-cum-peat mixture. Discretion should be paid to watering, and keeping the young plants warm, without allowing the air to get clammy. (Many books contain much information on saintpaulias, and there are works devoted entirely to them. They are treated here as minor occupants of the warm greenhouse.)

SAINTPAULIA IONANTHA. The African violet; this is the parent of the hundreds of forms and cultivars that are seen in shades of blue, pink or white. See catalogues for named kinds.

S. TONGWENSIS. This has slight botanical differences to above. There are two or three other species.

Sinningia (Gloxinia)

While new stocks of these are most often raised from seed, sown in pans in February in a temperature of not less than 21°C (70°F), the old tubers can be retained year after year, in time forming impressive specimens. They must be stored through the winter in their pots or pans, completely dry and as warm as the greenhouse normally is. They can be repotted in January, February or March, when a succession is wanted. Place each tuber in its fresh small pot, just below the compost surface, and merely moisten. When growth begins, give water more freely, and repot into larger pots when rootbound. They love the moist atmosphere, but must have light and should be put just under the glass roof. Shade later when the sun is very warm. Put small canes or sticks against them for supports. If there is a stove house, as well as a warm one, the stove should be their home before flower buds arrive, but they are all the better for the cooler one while in flower. When flowering ceases they must be given less water. When the tops wither it should completely stop.

Gloxinias sometime suffer from fungal rot at the base of the stem. As a precaution, I make a habit of using Cheshunt Compound upon them, as for seedlings. But correct watering is really the best means of prevention.

SINNINGIA CONCINNA. A species with purple flowers.

S. HELLERI. Has white flowers that are spotted red.

S. REGINA. Blue flowers with yellow markings.

S. SPECIOSA. This is one of the parents of the marvellous forms and cultivars of gloxinias bred by seedsmen and gardeners that give us brilliant art shades of white, blue, purple, rose, red and scarlet with trumpets up to 4 in. across, while the foliage is as tough and as soft as chamois leather. See the seed catalogues for named kinds.

There are natural variations of all these.

* * * *

KOHLERIA × 'LONGWOOD'. One of two or three dozen species that used to be listed in dictionaries and catalogues under various other generic names including *Isoloma*. It grows to about 3 ft. and has red flowers with white lobes that are spotted red. It needs similar treatment to the saintpaulia, being similarly everlastingly in flower.

RECHSTEINERIA CARDINALIS (syn. *Gesneria cardinalis*). This bears long tubular flowers with over-hanging lips, the colour being intense scarlet, lightly spotted purple. Season is summer and autumn.

SMITHIANTHA ZEBRINA (syn. *Naegelia zebrina*). Commonly called temple bells, this is the parent of hundreds of hybrids and cultivars offered under various strain names. The best way to start is to get a packet of first-class seed, from which will come a very mixed batch of plants giving flowers of red, orange or yellow shades, spotted with the same colours, while foliage will be bronzy and velvety, in keeping with the family.

Note. I have not included *Streptocarpus*, as it is a cool greenhouse subject and does not come within the scope of this book, although it is a member of the *Gesneriaceae*, and a very popular one too.

CHAPTER 15

Hippeastrums and other Amaryllids (Amaryllidaceae)

THE amaryllis family includes some breathtakingly beautiful flowers, as well as popular subjects such as the lovely daffodils and snowdrops. There are also nonbulbous species, but for the most part they bear flowers, build up the bulb again for the next season, and then have a rest until flowering time is once more due. There are comparatively few that need the warmth and atmosphere of a stove or warm house.

A good many come from South Africa, among them being *Cyrtanthus*, *Haemanthus*, *Hessea*, *Nerine*, *Strumaria* and *Vallota*, which are ideal in the cool house, for they then get protection without too much warmth.

There are a considerable number of bulbous amaryllids that are almost, if not completely, hardy. And there are many hybrids between the species, as well as garden cultivars, making them almost impossible to classify correctly.

Clivia

Clivias are evergreen, with broad, fan-shaped tufts of strong leathery leaves that in spring produce strong stems with umbels of brick-red flowers at the tops, quite exciting while they last, but for the rest of the year the plants are rather dull. The best place is the warm greenhouse. Time to repot, if they need it, is just before they want to grow in early spring. Then, if put into John Innes potting mixture or the standard one, they have little time to fret. After potting, or dividing and repotting, keep them syringed, and later, when they are developing nicely, give more water. It is at this time they love warmth and moisture. As with other bulbs, generally, they build up after the spring flowering, and then have a resting period, but in their case it is winter rather than summer. Never allow to get really dry, as they are evergreen.

Resenting root disturbance, they should not be repotted any more than necessary. But it does pay to scrape out the top compost each year and give a fresh dressing, as for hippeastrums.

CLIVIA MINIATA (syn. *Imantophyllum miniatum*). Flowers are formed in an umbel on a strong stem about 1 ft. high. Each flower is funnel shaped, of a definite hard scarlet colour with a throat like yellow ochre. There are several varieties in varying shades of the same colour. There are also a few other species.

Eucharis

In one stove orchid house under my care, we planted eucharis bulbs in a bed under the staging because it was the traditional way to grow this wonderful flower. They flourished. The overlapping leaves resembled some forms of hostas, and above them rose umbels of waxy white flowers, each about 3 in. across. They were fragrant.

Bulbs of considerable size have to be planted carefully, as the plants are evergreen. They have intermittent periods of flowering and of resting, and should always be dealt with during a brief resting period, if necessary. They are increased by offsets, it being necessary to sever them from the main bulbs. When leaves are forming, give plenty of water. When flowers are arriving, give a little fertiliser. When plants are resting, withhold water, but never allow to become dry.

They can be grown in pots or shallow tubs to considerable size. The bulbs increase and spread wherever there is room. John Innes compost or a standard loamy mixture suits them very well. They should be planted so that the top of each bulb is 1 in. below the surface.

They really love light, and should only be shaded when the sun is brilliant, which is in keeping with most other occupants of the warm greenhouse.

EUCHARIS CANDIDA. This has smaller flowers than the next species and sparse foliage. Its outstanding character is the prominent corona; in all other respects it is similar.

E. GRANDIFLORA (syn. *E. amazonica*). This has leaves on stalks that are about 8 in. long. Stems grow up through the dense foliage to a height of 2 ft. from the floor or pit, and at the top of each come four or five narcissus-like flowers, each fully 4 in. across, that are pleasingly fragrant. The corona is tinted green, and all the flowers droop gracefully. They keep well in water when cut.

E. G. FRAGRANS. Is regarded as a more fragrant variety. There are also other varieties of the species.

Hippeastrum

At one time amaryllis meant hippeastrum, but it is now the name of the belladonna lily. Most of the hippeastrum species are from South America, while the present cultivars are the outcome of much hybridising and cross-breeding, making them the most glorious of floral mongrels, with the possible exception of the modern lily cultivars, to which the amaryllids are regarded as inferior relatives.

If you buy good bulbs, such as are seen in the Keukenhof Gardens in Holland, they will arrive at any time of the winter. They should be planted singly in pots just large enough to allow an inch space between the pot side and bulb, about 2 in. down the inside from the rim.

For these the extra depth pots are best. 'Long Toms' are often used by shrub nurserymen, and are just right for this flower too. They allow a deep root run. Quite often we are told to plant them so that most of the bulb stands up above the soil. This may be for shallow pots, but it is better for future years if they can be buried to just above the bulge (see Fig. 11).

John Innes potting mixture is quite good. Or the standard 3 parts loam, 1 part peat or leafmould, and some sand with a dash of bonemeal, will do. See that drainage is perfect, this being essential. The soil should be pushed between the roots, if any, and be pressed down the sides to keep the bulb firmly in position.

The best place to put the potted bulbs is in the warm greenhouse, rather than the stove. Do not give any water for several days. Water given too soon sours the soil, but when top growth begins, roots will also start to form. Then give water freely.

Keep in a sunny position in the warm house until in full flower, when they can be put in the conservatory or wherever needed, if the temperature is reasonably equable.

After flowering, if no seed is wanted, cut off the flower stems to within an inch of the base. Foliage is now being produced quite freely, and it is most important to keep the plants well watered, the best position being the warm greenhouse. While the foliage is green, an occasional feed with Liquinure, or something similar, will help build up the bulb before it rests.

Some bulbs stay green longer than others. They tell you when a rest is necessary—the foliage yellows and shrivels. Then give less water, and by October practically none at all. In early spring, carefully scrape out the top inch or so from around the bulb, and fill with fresh compost. If the bulb is as

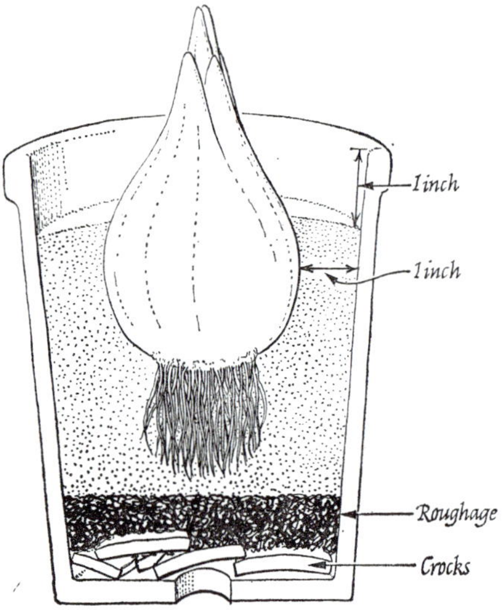

Fig. 11. Position for Hippeastrum bulb in pot.

large as the pot, put it into a slightly larger size, without disturbing the roots more than possible. Be careful not to give much water before growth starts again.

There is a prevailing idea that a cooler place, such as a cold frame, or even plunging the pots in an ash bed outdoors, is good in autumn. Actually, it can be more injurious than helpful. If given the correct watering and ventilation, hippeastrums will last for a decade and bear flowers every year, when kept in the warm house all the time.

The usual way of propagating true species, and special cultivars, is by offsets, which can be taken away quite freely at repotting time. New cultivars are most often raised from seed. They germinate quite easily, and the resulting plants will generally come into flower within three years. The main point is to keep them growing all the time for the first two years, and only after that allow a resting period, as for old plants.

Best time to sow seed is directly it is ripe in summer. Use the John Innes seed compost, or something earthy rather than peaty. Cover the seed lightly. Keep a temperature of around 16°C (60°F). In the autumn, transplant singly from the seedpan into small pots, and keep them watered and syringed throughout, being discreet regarding daylight and weather. Some growers plunge the pots into indoor beds of damp peat to keep the roots from drying out. As they grow, they are repotted into slightly larger sizes. Only if the two-year-olds lose their foliage for a while in the autumn, should they be kept dryish until they start to flower or produce new leaves.

HIPPEASTRUM EQUESTRE. This is rather variable, although there are several definite forms, but the chief one is green and red, and bears up to four flowers, each about 5 in. across, on a stiff stem about 18 in. high. It is one of the species from which modern cultivars have been evolved.

H. PARDINUM. Has rather a small bulb, but flowers of considerable width that are cream spotted crimson. There are variations.

H. VITTATUM. Flowers are white, striped red. This is stated to be another species much used for hybridising.

There are scores more true species for the collector, rather than the grower, of the gorgeous mongrels.

* * *

Apart from the hippeastrums, there are a few other members of the family that thrive under similar cultivation. Among them are the following:

ORCEOCHARIS CLIBRANII. This hybrid used to be grown in large gardens in precisely the same way as eucharis. It has an oval leaf, and umbels of pure white flowers that appear on tall stems in summer, more bell-shaped than those of the eucharis. It is an interesting bulbous plant.

SPREKELIA FORMOSISSIMA. Jacobean lily. Often offered in catalogues, and looks more like an orchid than an amaryllid. It has a definite lip and tenuous side petals.

STENOMESSON INCARNATUM (syn. *Coburgia incarnata*). The flowers, borne in umbels on stiff stems, are red and more elongated than those of hippeastrums. They are interesting novelties.

CHAPTER 16

Lilies (Liliaceae)

THE aristocratic lily family contains comparatively few genera that need, or are worthy of, a warm greenhouse.

Asparagus

The climbing greenhouse asparagus ferns are lilies. Though they are really cool greenhouse plants, they will thrive in the shady parts of the warm house. They need ordinary soil and should be potted afresh every spring. Roots can be divided, but the best way to increase is to raise from seed, which is freely available.

ASPARAGUS MEDEOLOIDES (syn. *Myrsiphyllum asparagoides*). The common name of this is smilax; (it should not be confused with the genus *Smilax*, which is a bit similar but far more difficult to grow), it is the trailing plant with the zigzag stems and elegant foliage used in floral decorations.

A. PLUMOSUS. Ideal for wreaths and bouquets, this has a very deep colour, and the sprays are flat and fine. Quite a good warm greenhouse plant, if you particularly like it.

A. SPRENGERI. A much more straggly and strong grower, of pale green, with prickly stems. Makes a good basket plant, but not refined enough for the choice collection.

Cordyline

These are closely allied to the dracaenas, and are often cross-named with them. Apparently the botanical differences are in the roots. Some look like refined yuccas, while others resemble polished aspidistras. Some are very interesting. Some are rather dull.

CORDYLINE HAAGEANA. Grows about 2 ft. tall and produces leaves that are about 6 in. long and 2 in. across.

C. PUMILIO. Makes a rounded head of long narrow leaves whose chief quality is its neatness.

C. TERMINALIS (syn. *Dracaena terminalis*). This is one of the best of the genus. Will grow to 6 ft. and has a bunched head of leaves some 2 ft. long. There are many variations, such as C. t. 'Ruby Variety'. The species has metallic green leaves with rose tints.

Dracaena

There are many species and varieties of the glowing dracaena that demand the stove, as well as several that tolerate the warm greenhouse. Some of the stove species produce remarkable variegated foliage in various shapes, and make first-class living-room plants, which can be used again and again if returned to the stove for brief spells to recuperate.

They grow after the manner of the India-rubber plant and sometimes lose their lower leaves. Air-layering is resorted to. An abrasion is made on the bark just below where the lowest complete leaf is growing. The leaf is cut off, and some sphagnum moss is wrapped round the spot, which is then covered with polythene or some other moisture-retaining material and sealed top and bottom. When it is known that roots have developed in the moss, the stem is severed through just below the new roots, and the new plant put into a suitably sized pot and placed in the stove where it gets plenty of moisture and warmth. The old stem of the plant that remains is then cut into pieces about 2 in. long, and just covered in a pan of sedge or wet moss peat before being put into the propagator. Each piece of this will also form the basis for a new plant, which is brought along by potting, and repotting as it grows. John Innes potting mixture-cum-peat suits it well. Besides this there is little to do beyond giving the normal stove treatment.

DRACAENA CONCINNA. This species has a main stem from which sprout, at the top, long, narrow, recurved leaves to form a sombre green and purple bunch.

D. GODSEFFIANA. A beautiful branching plant with bright green leaves that are flecked with creamy yellow spots.

D. GOLDIEANA. Leaves are quite different to *D. concinna*, being more aucuba shaped; the plant is as well balanced as a dwarf ficus. Leaves are yellowish green and beautifully marbled silvery green. A good specimen is most attractive.

D. PHRYNIOIDES. Leaves are 6 in. long, deep green, and

◀ *Peperomia caperata* 'Emerald Ripple'. A small representative of a large genus with decorative foliage.

◀ *Pilea cadierei*. An easy to grow member of the nettle family, with silvery leaves.

1 *Dipladenia* × *amabilis*
2 *Nepenthes* × *coccinea*
3 Bromeliads etc. in the new greenhouse at Wisley
4 Hippeastrum
5 *Spathiphyllum wallisii*
6 *Smithiantha* 'Elke'
7 Cattleya orchid
8 *Anthurium andreanum*
9 *Aechmea rhodocyanea*
10 *Eucharis grandiflora*
11 *Hibiscus rosa-sinensis*
12 *Tibouchina semidecandra*
13 *Gloriosa rothschildiana*

1

2

3

5

6

7

8

9

10

11

12

13

◀ *Platycerium bifurcatum.* The stag's-horn fern to hang upon the wall.

Ruellia macrantha. Christmas Pr a small shrub that bears purp flowers in winter.
▼

spotted yellow. It keeps dwarf and remains a good pot plant for a long time.

D. SANDERIANA. Sometimes listed in trade catalogues. Has leaves about 6 in. long that are striped white, or yellowish white.

There are a considerable number of varieties, or cultivars, in circulation with names like 'Gold Dust' and 'Firebrand', which give some indication of their colourings.

Gloriosa

These are the most colourful of the greenhouse lilies. In effect, they are tuberous-rooted climbers that go completely to rest in the autumn and winter, and start up again in spring. I have grown them in large pots by planting half a dozen of the long tuberous roots 2 in. deep in John Innes potting compost, watering them carefully at first and increasing the water and humidity as they develop, at the same time giving partial shade. Some people plant them singly in smaller pots. But with several in a large pot, they can be spiralled around the outside of four bamboo sticks inserted in the pot, to form a bower of pale green stems and leaves, and from side stalks come the beautiful flowers. Once the stems start to fade, water is gradually withheld until they rest completely and are then kept warm and dry in the pot until spring.

GLORIOSA ROTHSCHILDIANA. The stem and foliage are pea green. Flowers are lily-like, with narrow petals that fold back, the colour being ruby red with brilliant yellow edgings. The anthers are exposed and most prominent.

G. SIMPLEX. Like the fore-going, but the flowers are orange-scarlet with yellow markings.

G. SUPERBA. This is the strongest grower, and has rich orange flowers, relieved with red parts.

* * *

SANDERSONIA AURANTIACA. This genus and species is somewhat like gloriosa, but it makes a stiffer and more dwarf plant. Flowers are orange-coloured.

CHAPTER 17

The Melastomiads (Melastomataceae)

THE tibouchina is probably the best known genus of several marvellous stove and greenhouse subjects of an unfamiliar family. Everyone should know about them, even if there is insufficient room or they are difficult to obtain. They practically all grow quite easily under normal treatment in the warm house and need containers according to size. John Innes compost with peat is the best growing medium. The small ones start to flower at an early age, but the stronger and larger they are, the more profuse the flowers. Normally they are shortened after flowering. Keep the roots moderately dry in winter and use the syringe freely upon the foliage in summer, when not actually in flower. Give shade from the hottest sunshine.

MEDINILLA MAGNIFICA. A beautiful evergreen that grows about 4 ft. high and has bright green leaves 12 in. long and 7 in. broad. The rosy pink flowers arrive in branching panicles, while the unopened buds are also tinted pink. The stove is the best place for them. Prune back after flowering. Syringe while growing, but not when buds or flowers are forming in early summer. Increased by ripe cuttings in late spring. Keep shaded from hot sun. There are varieties.

MICONIA MAGNIFICA. Has reddish brown stems, and leaves that are 2 ft. long and a foot across, of a velvety green colour with striped veins and purplish hues. Makes a beautiful plant and stays dwarf. A non-loam compost is preferable.

MONOCHAETUM HARTWIGIANUM. This is a dwarf evergreen with smallish leaves and rosy purple flowers that arrive in great numbers during autumn and early winter. The best place is the warm greenhouse. Cuttings root readily in spring and the resulting plants will bear flowers by the end of the year. They are modest in their demands. The best compost is the John Innes compost-cum-peat.

TIBOUCHINA ELEGANS (syn. *Pleroma elegans*). This is a beautiful evergreen that is not so vigorous as the one below, with lovely glossy green leaves about 1½ in. long, and rich purple, spidery blossoms 2 in. across. It is best to put some loam in the potting mixture and keep it in the warm house. It will grow in the stove but may fail to produce flowers in too much heat. Some gardeners put them outdoors during August, in a spot where they are shaded from the hottest sunshine. The flowers generally arrive before midsummer.

T. SEMIDECANDRA FLORIBUNDA. The Brazilian spider flower is an evergreen shrub that can grow to 12 ft. It has lovely marked leaves and velvety purple flowers that come and go rather quickly in a steady succession over the weeks, through late summer, autumn and early winter. After having a short comparative rest (they should never be allowed to dry right out), the shoots can be pruned back to a neat shape during February. Propagation is by young growths taken off with heels. Sometimes this shrub is trained against a wall or a trellis, in the same way that wall shrubs are outdoors. But for moving around, the specimens in largish pots are most ornamental and useful.

Note. There are several other genera and numerous species in this family if one wishes to collect them.

CHAPTER 18

Orchids (Orchidaceae)

I PONDERED over whether to include orchids in this book, but knowing that there are some lovely members of that huge family which would enhance the beauty of the warm greenhouse and thrive in the company of the other plants, I have decided to give a brief introduction to a few of them.

Tropical orchids suffer from specialisation. If only they were regarded as being on a par with the rest of the indoor plants, they would be much more widely grown. The experts have ever put too much emphasis upon their personal knowledge and mystique, rather than advertising the orchids themselves and telling us how they can be grown without a lot of trouble.

At one time in my career, I was responsible for a collection of some 2,000 plants growing along with other subjects, such as have already been written about in previous chapters, in six greenhouses of varying temperatures. Among them were some that grew easily—in fact, they did better than the other plants. Such as these I mention here, bearing in mind that the stove and warm greenhouse conditions as described must be maintained, the orchids being secondary to the truly ornamental subjects.

The best way to know how to treat them is to be conversant with their simple characteristics and habits. Once we grasp the fact that they are different from normal vegetation, although less difficult to manage, we are on the road to success.

No doubt, experts could take me up on many details, but they will have to agree, I think, upon my generalisations and broad principles of cultivation.

There are two main categories of orchids. By far the larger is the epiphytic one, involving all those that lodge, rather than live, upon other vegetation, much as do many of the bromeliads and gesneriads dealt with in the earlier chapters. They rely upon wind to blow the seed about, tropical rains to sustain them, and a density of humidity, loaded with vapour and the fumes of decayed life, or converted organisms, as well as falling leaves from above, and even bird's droppings. They have both anchor, and also aerial roots that grow, cling, or grope for food from the air.

There are also orchids that live upon the earth and do not have aerial roots. These are terrestrial. The cypripediums comprise the main group in this category. Their habit is to form clumps of evergreen leaves in quite a different manner to the majority of epiphytes. Nevertheless, there are also a few species of these that do live an epiphytic life. All love to grow in soil full of dead vegetation or humus. It is just a matter of how high they are above solid ground, which is never very much.

The vast majority of epiphytes have built-in conservation reservoirs in the form of very fat stems, often referred to as pseudobulbs. These vary in shape according to their kind. The normal habit is to produce one or more of these each season, in keeping with the size of the plant. Upon each pseudobulb come the leaves. Often these are thick and are merely extensions of the thick stem below. Sometimes they are grassy. The flowers arrive after the resting period of the newest pseudobulb has been completed.

Some arrive through a sheath at the base of the leaves, and others come at the base of the pseudobulbs. In a way, they arrive much like the daffodils do. They build up new pseudobulbs, and then have a resting period before producing the flowers. Having completed the flowering, they often have another brief resting period before beginning the cycle again.

The pseudobulbs are produced on rhizomes, much like the growths are on the German bearded irises of the borders. Sometimes these rhizomes are pronounced, and sometimes they are almost indiscernible.

Having produced flowers, the bulbs then become the reservoirs to feed the new front ones in times of drought. Some retain their leaves. Some are deciduous.

When buying plants of species that form pseudobulbs, it is possible to judge the healthiness of the individual. A good plant will have a good strong pseudobulb at the growing end, while the predecessors will be slightly smaller according to their ages. But should these pseudobulbs be in reverse order, i.e. the weakest ones at the growing end, or ends, the plants are definitely 'going back'. They rarely turn to normal healthiness once they get into this decline.

Raising new plants from seed has become a highly scientific business among the large specialist nurserymen and private growers, making it possible to obtain new vigorous seedlings at reasonable prices. That form of propagation is out of my realm, but I have increased stocks of many plants by division, which is done at potting time. We will consider some of the ways that it is done when we deal with the plants.

Potting

While it is interesting, and more expert, to grow some plants on blocks of wood, or bark, all those we favour can be sustained in baskets, pans or pots. Some, such as the oncidiums and

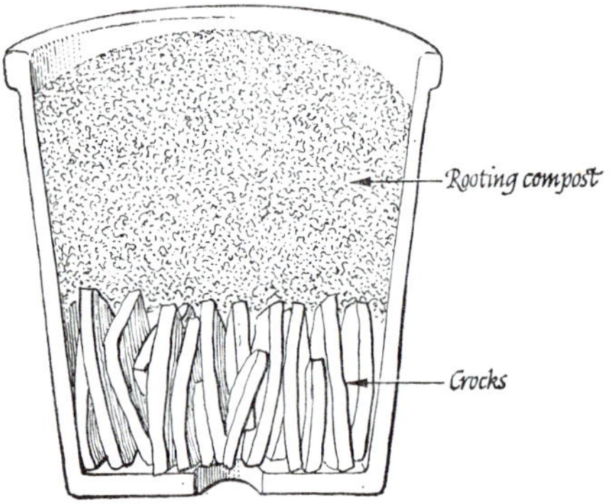

Fig. 12. Crocks and compost contents for orchid pots.

miltonias, need very small pots, while larger plants want those of considerable size. But they should never be planted very deeply. They grow best when the necks are well above the pot rim levels, the materials in which they are growing forming nice rounded low mounds.

Clay pots are preferable. They should be half filled with broken tiles or potsherds stood endways up (see Fig. 12). Both pots and potsherds should be sterilised before planting begins.

The compost consists of osmunda fibre and sphagnum moss. The proportion is generally 2 parts of fibre to 1 of moss. Both are chopped up finely and mixed together. If live sphagnum can be gathered, as has been the case with me, it is far better than having to buy that too. However, if this is difficult, special orchid compost can be obtained from firms like Bentley Ltd (see Chapter 26) or some orchid firms will be pleased to supply.

In more recent times, the specialist growers have been using other materials. They include dead beech leaves and dead pine bark. If you can get such materials, either collected from the woods or bought, a good compost could consist of osmunda fibre, sphagnum moss, beech leaves and pine bark. Or either the leaves or bark can be omitted. Osmunda is expensive. But I still think that it and sphagnum should never be left out of a mixture for epiphytic orchids. Whatever is used should be roughly chopped up and mixed together and well dampened. In good condition it feels soft and moist, but never soddenly wet, when rubbed between the fingers.

Potting an orchid is an art. You have to encase the anchor roots in the given mixture without damaging them, yet making them firm to the extent that you could not lift one by a leaf without lifting the pot as well. You put a nest of the stuff under the base of the plant, and sit that on the edged crocks in the container. You then push in more mixture down the sides and fix it with the aid of a round peg, as thick as the little finger. By pushing it down against the pot surface, and then pressing it outwards so that the pointed end goes inwards under the plant, you weave the mixture until all is tight and neat.

Watering

The roots of the plants are given water according to the time of year and their condition. In the summer, the sphagnum gives one a clue. As long as it remains green that is sufficient. When it starts to shrivel, water is poured into the pot, unless the orchids are resting.

The main watering is the sprinkling around from a fine-rosed can, the paths and the top of the under-staging getting the most. In winter, the heat of the water-pipes causes the shingle or chippings on the under-staging to dry out. Only when there is a threat of its being really dry is water spilled around. This could be needed once or twice daily. During spring and summer, the artificial heat, when on still, makes a lot of water-sprinkling necessary. In the height of summer, when we rely upon natural heat, with no help from the boilers, the water upon an under-staging has to be fairly well dried out before any more is given. Sometimes the nights are cool and the mornings dull. Then it is advisable to withhold any watering until the temperature rises to 16°C (60°F) (one of my assistants used to gauge the correct time by feeling an orchid leaf—it had to be dryish instead of clammy).

When the sun rises early the blinds are pulled down. At

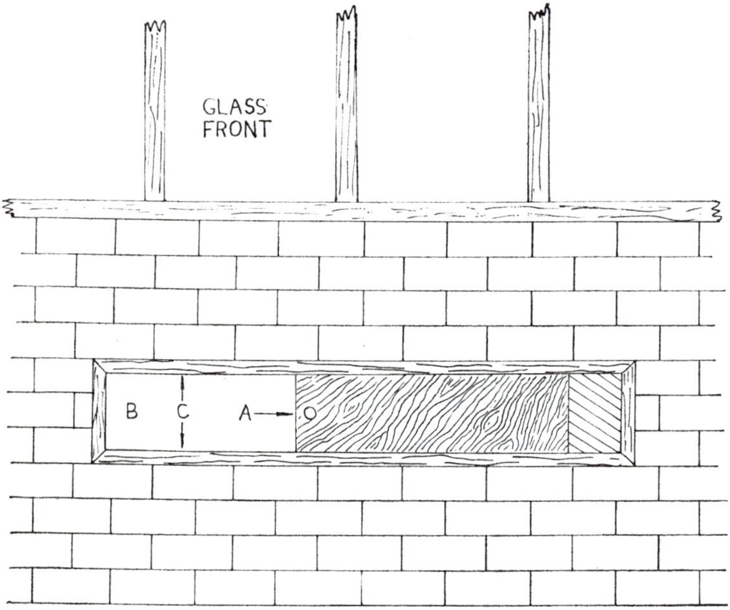

Fig. 13. Side wall ventilation for orchids.
A—Wooden shutter in slide guides—C
B—Ventilation gap in brickwork.

18°C (65°F) or thereabouts, the sliding door of the box ventilator in the brickwork is pushed open. In very hot spells, during the longest days, the door is left open. And by damping down, the temperature is kept to about 25°C (77°F).

Our greenhouses containing that largish collection had not been constructed especially for orchids, but were converted to suit their needs. The one we called the 'cattleya house', the general management of which has been described, was formerly a plant house before its conversion. It had wooden slatted staging around the sides and stand in the centre, enclosed by the paths. It also had top and side ventilators which had had the gear removed, so that they could not be easily opened.

In the place of these, slots were made in the brickwork, with sliding lids, that enabled them to be opened to any degree. There was no possible strong draught, and in winter the air passed over the warm pipes before it entered the house, no doubt escaping through the inevitable crevices in the glass-work, and when anyone entered through the door. This avoided direct draughts (see Fig. 13).

Under the slatted staging was fixed a more solid one to carry a layer of shingle. Incidentally, when this became dirty through dripping from the pots, it was shovelled off and placed in a fine sieve and washed by steeping it in a water tank and swirling about, before being put back again straight from the sieve.

There are hundreds, if not thousands, of species of orchids that will live happily in a heated greenhouse, along with a collection of other tropical plants needing the conditions I have frequently mentioned. These are: the comparatively brief resting period of winter; the light shading of summer; the spring potting time, when essential, using a compost that suits all epiphytes. Let me introduce just a few that I think are most desirable. And some that are not.

Calanthe

At one time calanthes were grown in larger gardens where no other orchids were included in the plant collections, because they responded to the usual greenhouse routine treatments.

There are two distinct types—the evergreen species, and those that completely lose their foliage in winter. The latter were the ones most cultivated.

Being terrestrial, they need loam added to the osmunda fibre and sphagnum (also beech leaves and pine bark if used). Some gardeners successfully grew them in a compost of 4 parts roughish fibrous loam, 1 part chopped sphagnum, 1 part leaf-mould or sedge peat, and some potting sand if the loam was on the heavy side.

They have to be repotted every spring. The bulbs, looking rather like those of garlic, with a somewhat silvery hue, should be planted firmly in the compost, but only a little of the lower part buried, as with hippeastrums. Those bulbs of the cluster that are more than two years old are broken away from the newer ones and are planted singly or in pairs, in other pots, in this way increasing the stock. These will, most times, produce new bulbs, and eventually flower spikes that grow from the base.

The leaves form on top of the bare bulbs, and look a bit like those of hostas. While growing through the summer they are given plenty of water, but this only starts when they show signs of activity after the potting. By September, water is given more sparingly and, when the leaves begin to yellow, is reduced to merely keeping them damp.

At this time, when leaves are in the dropping-off stage, flower spikes will grow to 2 ft. high and bear the butterfly-like blossom along the top half. Once they fade in the dead of winter, the bulbs are kept quite dry, in a warm temperature, until potting time comes once more in March. They respond to a little liquid fertiliser during the growing period.

There are many varieties, one of the commonest species being *Calanthe vestita,* whose white petals form a halo over a prominent lip of peculiar shape that seems to have been rouged. While they do cheer up the greenhouse, or are used for cut flowers, at a time of scarcity there are many more orchids much more exciting.

Cattleya

Many of the cattleyas are easier to grow than they look. These come from the same parts of America as many of the bromeliads. The warm greenhouse seems to suit them better than the warmer stove, and they enjoy the conditions that normally prevail for other plants.

While the species do vary in shape of pseudobulbs, flowers and leafage, those that are most amenable have a rhizome, from which rises a series of swollen stems surmounted by one or two long, leathery, wide leaves. From the point where the leaves grow from the stem a sheath is formed, and from that emerges a cluster of the marvellous flowers of extraordinary colour, that last a very long time.

Some species and varieties flower in autumn and winter. They should be transplanted in spring—if they need it. Those that bloom in spring should be repotted just after the flowers have faded. But this need not be done every season. As long as there is enough room in the pot or pan, they live on happily.

Perhaps, if the compost gets tired, and the sphagnum content dies, a little replacement on top of the pot or pan can be carried out.

They are increased by division. A good plant will have several of the thick stems (pseudobulbs), consisting of those that have previously flowered and one or, maybe, two new growths for future development. Always see that there are three old back bulbs behind the latest new one. When it is intended to divide, the rhizome should be cut halfway between the third and fourth pseudobulb one year before it is actually completely severed and taken away.

The piece with the new growth should then be repotted, leaving the older back part in its original pot, merely filling the gap with compost. Both the severed piece and the old, now in their two pots, are given a thorough watering, but are not given any more until they show signs of growing again. Once there is renewed life the roots are kept moist, and the air is filled with humidity, as is normal in summer in a warm greenhouse.

In autumn comes a resting period. The plants are kept drier. The sun can shine upon them from Michaelmas onwards through the winter. Cattleyas, like so many of the other tropical plants, love light, and even sunshine, except in summer. I always try to imagine them growing in forests where they get the shade of tree leafage in summer, but in winter those trees are bare of foliage.

One of the most successful species we grew was *Cattleya skinneri*. It had a spike of nine or ten flowers, each 3 in. across, of a brilliant rosy purple colour. The flowering season was around Easter. I think it liked our treatment because we normally potted, or repotted, the main collection during the the first half of May, the time which most suits cattleyas. It was impossible to give all the other orchids their correct potting season.

It is impracticable to give more than an inkling of the many hundreds of species and varieties but, generally, just to augment the other plants, I think that the ones that bloom in the spring, rather than those that arrive in autumn and winter, are the most likely to flourish.

One is *C. amethystoglossa*, which bears about half a dozen flowers of rosy hue, dotted purple, with a deeper purple lip.

Each is about 3 in. across, and the cluster arrives during late spring. There are several varieties of this species, including one that has white flowers.

Coelogyne

Here is a large genus of orchids that varies in temperature needs. Several will flourish in the warm greenhouse if given the normal routine treatment. They are best grown in baskets or pans, the spikes of flowers growing from the clumps of pseudobulbs and hanging over gracefully as though weakened specimens of weeping cymbidiums. We had a specimen of one species, *Coelogyne pandurata*, that was most attractive. The flowers were a mixture of bright green and soft black, of extraordinary shape and wonderful fragrance. They hung upon a curved drooping scape about 2 ft. long. Each was about 3 in. across, and there were nine or ten in number.

The names are interesting. *Coelogyne* means, literally, a hollow female, while *panduratus* means fiddle-shaped, this aptly describing the formation of the green and black lip.

Cymbidium

These are probably the most numerous, and most grown, of orchids in more modern times. They are florist's delights, on account of their long, strong stems, studded with bold butterfly flowers in a thousand different hues, from cream and green to deep pink and chocolate-brown. For all their desirability, I do not regard them as good companions for the mixed stove or warm greenhouse. They need lots of room, good specimens filling 10-in. pots, and treatment that is not in accord with the general conditions of other subjects.

They need fresh air, and a cool temperature in the autumn, when they are allowed to become almost dry. When the flower spikes begin to show, they are given a thorough soaking, by plunging each pot in a tank of water and then placing them in moderate warmth, keeping moist until the flowers fade. When potting time comes in March, not every plant will need it. The mixture should be as for calanthes. After the potting, there should be one watering, until fresh growth is obvious, and then plenty of moisture applied until August. The unpotted ones should be given frequent doses of liquid fertiliser. Propagation is by planting the old, detached pseudobulbs in spring. They should never be covered, but merely fixed so that they nestle on the compost surface.

Cypripedium

Slipper orchid, moccasin flower and cypripedium are all familiar names for the greenhouse orchids that comprise what should really be called the genus *Paphiopedilum*. There are vast assortments of them, the best known needing only the warmth and conditions of the cool greenhouse. There are also some really exciting species and varieties that flourish in the stove or warm greenhouse, and are very much worth growing when they can be obtained. The chief qualities are the beauty and the longevity of the individual blooms with the familiar pouches, as well as some extraordinary forms and markings.

We grew several plants of *Cypripedium* × *maudiae* in the cattleya house that were quite distinct from the 400 plants of the cooler section in their own house. These have beautifully mottled foliage, which makes them perpetually attractive. The flowers are a mixture of green, pale green and white, in a neat pattern, and arrive at different times of year, each on a slender stem, and lasting for many days. They need the compost recommended for epiphytes. Spring is the best time to repot, and most of them should have it done annually. They should be put into a slightly larger size, or be divided, as there are no pseudobulbs, nor aerial roots, and so they rely entirely upon the food in the rooting medium. Happily, they are not exacting.

We also had in the cattleya house a plant named *Selenipedium grande*, whose official title is probably *Phragmipedilum caudatum*, or *Cypripedium caudatum*, that had the usual slipper orchid face, while the side petals bore ear-drops or twisted ribbons hanging down to over 2 ft. long. We placed the plant in its pot on top of a stand, so that these petal extensions were just clear of the staging. The plant grew happily and was adored by all who saw it.

Dendrobium

There are over a thousand sorts of dendrobiums. They are exquisite flowers, often most fragrant, although not always pleasant. I merely mention them because we grew a lot in a particular house. They like more sunshine in autumn, and dryness in winter, and more heat and moisture than is usual in summer. If they can be given the more extreme treatments, then only are they worth consideration.

Laelia

Closely related to the cattleyas are laelias. Generally, they are less impressive, having more graceful blossoms, as well as much narrower petals and more elongated stems (pseudobulbs). A beautiful representative is *Laelia harpophylla*, which produces a spray of up to a dozen orange-scarlet blossoms looking like tropical butterflies with recurved tongues. These appear in spring and stay on the plants for several weeks. They are less satisfactory for cutting than cattleyas.

Because of their similarity, the cattleyas and laelias have been much hybridised, both in their native Brazil and by artificial means. We had a large number in that cattleya house that were no trouble at all.

Miltonia

Miltonias are lovely, smallish orchids that grow well in the stove or warm greenhouse, and bear blossoms somewhat the shape and size of pansies, but much more distinctively crenellated. They need small pots and grow about 9 in. high. The pseudobulbs are squat, and often enveloped by small leaves, while the flower scapes start from the base, generally with one solitary flower at the top, or there may be two or three on a fork. Flowering season is autumn.

They are definitely epiphytes, and need that sort of treatment. The usual summer moisture and the normal winter moderation in this respect is what they need. Any necessary repotting should be done in spring. They are propagated by division at that time.

They are related to the wonderful odontoglossums, that do not succeed in the warm or the stove greenhouse. Odontoglossums and their hybrids need to be kept too equably cool throughout the year.

Oncidium

Apart from the gorgeous cattleyas, I regard the species *Oncidium varicosum* to be one of the most desirable orchids to grow. It will thrive in the warm greenhouse, and exist in a stove temperature. The usual requirements for epiphytes regarding summer moisture and winter moderation are suitable. Potting time is spring.

Small pots are needed. One of the bothers is that the plant can become very top-heavy during the flowering period, and one way to steady it is to place its pot inside a larger one during that time.

This species has egg-shaped pseudobulbs, each with a pair of long, curly leaves at the summit. About four dozen blossoms are borne, on a branched wiry scape up to 3 ft. tall that suggest a cascade of golden stars, as from a rocket. The large, yellow lips are headed by a crest of dull red and brown. The season is November, and the flowers hang until they endanger the life of the parent. It is best to remove them directly they begin to fade.

Oncidium crispum is somewhat like the last-named, but bears its flowers in autumn. If anyone wishes to collect the species and the varieties there are around half a thousand. Many come from Brazil, as do these two, and thrive along with bromeliads.

Vanda

Vandas come from south-east Asia, and some grow well in the warm greenhouse. One species we had was the famous Blue Orchid—*Vanda caerulea.* It stood on the topmost shelf of the central stage, by the *Cattleya skinneri* plants, and produced each late autumn a lovely spray of seven or eight flowers, each about 4 in. across.

There are many varieties of *Vanda caerulea.* There are also many other species needing varying temperatures and treatments that may, or may not, like stove or warm greenhouse conditions. Many of them need much warmth. Generally, the usual conditions I have frequently referred to, for bromeliads and other epiphytes, will suit the majority. They need to be kept moist, but not too much water should be given in winter.

Vandas do not have apparent pseudobulbs, but send up a strong stem, from which firm leaves grow tiered criss-cross style, somewhat like leeks. At the top the flower stem appears from the axil of a leaf, while at the bottom grow aerial roots. One way to propagate them is to cut off the stem just below the lowest leaf, if there is a space, and put this top into the normal compost and keep moist. The part left may then make a fresh stem. But in order to be successful, there must be aerial roots at the base of the piece that is removed, and those roots must be woven very carefully into the compost. If there are also suckers coming from old plants, they are better for growing, but they must be allowed to get to a fair size before the amputations.

Note. As stated at the beginning of this chapter, orchids are very much specialist's pets. There is one consolation, that it is doubtful whether there is an all embracing expert, so vast are the members of this family. Those I have mentioned are but a few that are better known than most of the rest. There must be others that would flourish in a mixed stove, and some that would grow well in a warm greenhouse.

We are informed, that there are over 1,500 known species in 600 genera, which is almost beyond our comprehension. But that is no reason why, because of that, we should leave their cultivation to the specialist. In fact, it is to the advantage of the commercial orchid growers and collectors if we regard them as ordinary occupants of the greenhouses, rather than as the prohibited and unattainable.

CHAPTER 19

Palms (Palmae)

ALTHOUGH palms are elegant and beautiful, and grow well in the warmth and atmosphere of the stove or warm greenhouse, they are not considered, as a whole, really good subjects for mixing harmoniously with the other occupants. They are too aggressive. However, there are some less overwhelming than the rest which, where there is room enough, will add to that exotic effect that should always prevail in such homes for plants.

There are two main types of palms. There are those that have long feathered leaves, either herring-boned or undivided. Likewise, there are those with fan-shaped leaves, either divided or solid. Some of both types develop trunks, caudices or stems (call them what you prefer), and others are content to renew their leafage from ground level throughout their lives. Another subdivision is of those that bear spines and those which have no such appendages.

It is interesting to see how palms find ways and means of preserving and assuring themselves of perpetuity. The main way is producing seed, although the flowers are insignificant and form within spathes, which saves them from being beaten about by gales. Often they are hard-skinned nuts, probably the biggest seed in existence being the coconut. Then there are dates, which can be eaten and the stones carried away in animal intestines. We can raise new plants by their seeds but, whatever the kind, they have to be imported, because of pollinating and ripening difficulties in this country.

The tree palms are competitive, always striving to get their heads above their neighbours. They do this slowly, taking many years to attain the ultimate. The new leaves are formed on top of the trunk, one above the other, their bases increasing the height. A hard circular wall is built up, while the middle becomes filled with decaying matter that helps to retain moisture, and to some extent gives sustenance. The ground palms are content to grow in shade, and defend themselves by spreading the leaves far and wide, suppressing all below them.

The palms with spines cling to other vegetation, inflicting punishment upon their victims, and then bask contentedly in the sunny humidity. Others persist by sending up suckers. The chief thing about all this is that by understanding them we grow them more confidently. When the time comes, reluctantly we must consign the decrepit to the bonfire.

However, if we keep them in subjection, they remain manageable for many years. They must never be pruned, although it may be advisable to remove lower decaying leaves just for looks. We grow them in pots, tubs or containers that restrict the roots.

The best rooting medium is, in my opinion, John Innes potting compost (see Chapter 1). However, a lighter one is preferred for raising the plants from seed. One of the proprietary, no-soil composts will do for that. According to the size of seed, so should they be covered. For example, the kentia palm (*Howea*) seed should be buried 1 in. deep. The procedure is much the same as for tomatoes. Start them in brisk heat. Keep them watered. After they come up in the pan, or whatever is used, they must be potted off singly into 4-in. sized pots. When the roots have spread right through the soil, they should be put into slightly larger sizes. On no account is it advisable to give more than slightly larger sized pots at any one time.

Palm pots should have sufficient drainage to allow free water to escape. Cover the crocks with a little rough peat before putting in compost. When repotting is necessary, it is generally found that the strong roots have enveloped the crocks. If they can be straightened out and the crocks removed without much difficulty, that should be done, but with as little damage as possible. Any damaged roots should be trimmed away.

As the fresh pot should only be slightly larger than the current one, not a lot of crockage can be put in the new pot. It takes very little compost to fill the gap between the side of the fresh pot and the root ball, which is slightly roughed up, but never broken. Having wriggled the ball of roots so that it settles firmly on a small heap of compost placed over the crocks, the gap between pot and ball is filled. I found that an ordinary, wooden, 1-ft. school ruler was most useful to force the compost down this gap, making it as firm as possible. It is better than a round peg.

Established plants need only be repotted once every three or four years. However, the top surface in the palm pots can be scraped out and some new put in its place.

They need frequent feeding during the growing season. One or other of the proprietary liquid fertilisers will be good. Actually, I believe in changing the brand when a new bottle is necessary. I think that plants, although static, like a change of diet as much as we do.

Because the roots are restricted, they need lots of water in the summer. If the drainage is good you can hardly give them too much too often. They love to be syringed. Should they ever really dry out, then the pots should be put in a pail or tank of water for an hour or so to become thoroughly soaked and restored to the normal state of being moist throughout. Little is needed in winter, but practically all palms should never, at any time, suffer for lack of moisture at the roots.

One of the old chores was sponging the leaves. That does not appeal any more. Nevertheless, when it is carried out, it freshens up the palms and clears away any pests that decide to live upon them. If a dash of washing-up liquid is added to a bucket of very lukewarm water, the sponging will impart a lovely shine and enhance the elegance of the tribe.

We must now mention a few of the most suitable, or useful members of the family that can be grown in a stove or warm greenhouse without being incongruous in the company of the other occupants.

Kentia Palms

This is the general name by which two species of very popular palms have been known for generations. They are not kentias, however, but howeas. They were once included in a large genus, *Kentia*, which has since been sorted out into other genera.

To caterers, hotel managers, cinema usherettes—all and sundry, where these graceful palms are to be seen—they will ever be kentias.

They most often stand 6 ft. or more high, and fan out to make a fountain effect. They are subjected to all sorts of bad treatment and indignities, such as being wrapped around with newspaper and packed in cold vans, or handled roughly by being clasped around the stems just above the pots, and thrown about. Or when party visitors wish to lose the whisky foisted upon them, they find palm pots good means of averting embarrassment. The palms survive all these indignities.

ACANTHOPHOENIX CRINITA (syn. *Areca crinita, Calamus dealbatus*). The prickly date palm is often recommended for greenhouse cultivation. It has elegant, arching, feathery leaves of a rather pale green colour. It is armed with spines, and looks sickly if given insufficient heat.

A. ALEXANDRAE (syn. *Ptychosperma alexandrae*). This has a slight main stem from which grow rather fragile, broad, feathered leaves, the leaflets also drooping gracefully to suggest a beautiful plume. It comes from Australia, and will flourish in the warm greenhouse if given normal treatment. An attractive palm both while young and aged.

CALAMUS CILIARIS. The Rattan palm belongs to a genus of climbing palms, but makes a good pot plant for several years. Has thin leaves about 2 ft. long, bearing many leaflets that resemble feathers. Needs the stove to be good.

CYCAS CIRCINNALIS. The sago palm has a slim, yet strong, stem, whose exterior suggests crocodile skin. The leaves are stiff and erect, quite 6 ft. long or more, and are definitely fern-like in appearance. This palm makes an exquisitely beautiful specimen for several years and is a wonderful canopy for other plants in the stove.

DAEMONOROPS PALEMBANICUS. Closely related to *Calamus*, the majority of the species are strong climbers. This is more erect and makes a good pot plant while young. Has arched leaves with long leaflets of a reddish hue when small. Stalks are spiny.

HOWEA BELMOREANA (syn. *Kentia belmoreana*). The curved palm is the one with the arching stems, and prettily curved, long leaves.

H. FORSTERIANA (syn. *Kentia forsteriana*). The flat palm has stems stronger than those of the above species, and the leaves are less curved. These are the only differences between the species.

LATANIA VERSCHAFFELTII. Somewhat like *Verschaffeltia* but with a stiffer and taller stem. The leaves are round, and are broken at the edges. The colour is bright green. The problem is that this and other species need considerable room before they can be seen at their beautiful best.

LIVISTONA CHINENSIS. This is really a cool-house plant, but represents the fan-shaped palms, and is easy to grow as a house-plant. But it grows tall quickly, and is prone to pests. The species *L. rotundifolia* is often raised and sold by florists, the leaves being almost round and very attractive, particularly when the plant is young. The ultimate height can be 80 ft.

PHOENIX ROEBELINII. Rather like Syagrus, but is one of the dates. It is quite as handsome, though far less bright. Has a short bulbous stem and shortish leaves, these being decorated with rather long leaflets. There is a proneness to send up suckers, which must be removed.

ROYSTONEA REGIA (syn. *Oreodoxa regia*). The Cuban royal palm, when fully grown, is the most elegant of all the feathered palms. The trunk is then about 30 ft. high and rises up like a pure marble column, bulging somewhat in the middle. The skin is smooth and milk-white, while the leaves growing from the top are 10 ft. long and the leaflets 2 in. in length. The young plants are elegant when small, but never attain the glorious maturity until too large for the greenhouse.

SYAGRUS WEDDELIANA (syn. *Cocos weddeliana*). The Weddell or cocos palm is a miniature of the palm world, and considered to be the most adaptable in the warm greenhouse, or for putting upon the dining table. The stems are clothed with black fibre; the leaves grow up to a yard long in lovely arching formation, while the leaflets are a lovely bright green. It is related to the coconuts.

VERSCHAFFELTIA SPLENDIDA. This remarkable palm has a spadix (stem) some 3 ft. tall, rather slender and slightly bulging in the middle, and from the bark grow masses of spiny bristles. At the top are short-stalked, undivided leaves that are only serrated at the edges. They suggest elephant's ears.

CHAPTER 20

Peperomias (Piperaceae)

IF asked to name a house-plant when you have no idea of what it is, the chances are that it could be a peperomia, so numerous and variable are the species and varieties. This may be wrong, of course, unless you are conversant with the other genera and species. There are about a dozen different species grown in homes, while there are more that need a stove or warm greenhouse.

All are small plants needing small pots, a good compost mixture being the usual John Innes compost-cum-peat. Potting time is whenever the roots are overcrowded. These plants need less water than most others. If given too much they rot. This is particularly so if water rests on the crown of the plants. During winter they need only be dampened from time to time. They need light without bright sunshine.

They can be propagated from seed, or by leaf cuttings as described for saintpaulias, or by small pieces of shoots inserted in sandy peat. Do not put in a closed-in propagator filled with too much moisture, or they damp off. Some are affected by virus disease, in the form of zonal marks that causes disfigurement. The only way to control it is to destroy afflicted stocks.

PEPEROMIA CAPERATA. The overlay of leaves one above the other in close formation gives a busby effect, with a plume of cream mouse-tails at the top. The leaves are heart-shaped and corrugated dark green.

P. HEDERAEFOLIA. This forms a neat hummock of medium-sized, fleshy leaves of an attractive grey colour, the surface being raised among a pattern of veins. There is a silver overtone.

P. MAGNOLIAEFOLIA. Is an erect and lovely plant, larger than most others in the genus. The leaves are oval and spoon-shaped, of a dark olive green shade. There are varieties, such as *variegata* with cream tints and 'Green Gold', the names being descriptive.

P. MARMORATA. Somewhat like *P. caperata*. The leaves are

fleshy and of a rich shining green, marbled or marked white somewhat like cyclamen. The flowers are also mouse-tails, or catkins, these standing more erect.

P. OBTUSIFOLIA. Grows about 1 ft. high. The leaves are spoon shaped and have red margins. It likes warmth and moisture.

P. SANDERSII (syn. *P. argyreia*, *P. verschaffeltii*). Sometimes called the 'watermelon begonia', this is probably the most grown of any. It has a neat bushy habit, something after the manner of a saintpaulia. The leaves are thick and fleshy, heart-shaped, and the bright white flesh is decorated with radiating lines of silver to suggest a melon, if looked at from a distance. There are experts who regard *P. argyreia* as being distinct from *P. sandersii*, contending that the former has more prominent silver stripes.

Note. There are many cultivars, such as *P. caperata* 'Emerald Ripple' which has olive green leaves, which are quilted and ridged.

CHAPTER 21

The Rubiads (Rubiaceae)

THE famed gardenia belongs to this unexpectedly large family of plants that needs the stove or warm greenhouse in which to thrive. They are mostly grown for their floral qualities, although they have bright evergreen foliage, making them presentable subjects in their own way throughout the year. The general cultivation needed is heat and moisture during summer, and warmth and less water in winter. Most prefer some shade from the hot sun, but like as much light as is obtainable in winter. They love being syringed when growing.

They are propagated by firm shoots taken off in spring and summer and rooted in the warm propagator. After being potted in the usual John Innes compost-cum-peat, they need to be fed when mature.

This family is prone to pests more than most, and a close watch has to be kept for mealy bug, scale, greenfly and red spider mite. However, if introduced free of such parasites, they should be all right. One other bother is canker of the stems at soil level. When it does occur it is best to take care to take cuttings only from plants free of this trouble.

Bouvardia

Probably the longest flowering of the family, producing a continuous succession of flowers for cutting, while the specimen plants themselves are quite presentable. The originals came from Mexico, which indicates that they need warm greenhouse conditions. However, the best flowers come from plants that are kept in a cool house during spring and summer. Sometimes they are placed outdoors for a few weeks during July and August. Practically the only time they need the fullest degree of heat is when being propagated. After flowering in autumn and winter, they are stored in the greenhouse, given less water, and in time will sag. Then they are cut back severely, so that new shoots grow up from the base. While these are still short, they are removed, trimmed, inserted in a sandy compost mixture and given the usual hot, moist conditions of the propagator. Except while

Puya berteroniana. An unusual type of pineapple, but there are many species and cultivars.

◀ *Scindapsus aureus*. Evergreen climber with glaucous leaves. Belongs tc arum family. Sometimes known as the Ceylon pepper.

Stephanotis floribunda. Clustered wax flower—a climbing shrub whose flowers are very fragrant.

very small, they thrive in John Innes compost-cum-peat. They need plenty of light, and moisture when growing freely, but should be shaded from hot sun.

BOUVARDIA HUMBOLDTII CORYMBIFLORA. Has bunches of sweetly scented flowers, larger than those of the following, from which it is possibly derived as a hybrid.

B. LONGIFLORA. Has large, white, scented flowers, and is of rather dwarf habit.

B. TRIPHYLLA. A species that bears triple leaves and scarlet flowers.

There are varieties and cultivars, some in cultivation for many years whose names suggest they are species. 'Alfred Neuner' has pure white double flowers; 'Dazzler' is a brilliant scarlet; 'President Garfield' is pink, with double flowers.

Gardenia

As stated, these are the most renowned of the family, possibly because of the name, which is in honour of a Doctor Garden who lived 200 years ago. The many species come from various parts of the world, including China and South Africa. Special points in their cultivation are to give plenty of water when growing, and to syringe well during bright afternoons. Some are naturally bushy, but those that are at all spindly should have the growing tips pinched out. These grow better than most subjects in a no-soil, coarse peat compost. By careful management, such as encouraging them to rest by slightly reducing heat and water, or stimulating them by syringing and giving more heat, they can be induced to flower at different times. However, one of the characteristic habits is to produce flowers intermittently, rather than in one grand display, and for that reason they are hardly ever magnificent, although always beautiful. The flowers are used in bouquets, vases and, of course, in making the proverbial buttonholes. Young plants are the most rewarding, and stock should be constantly renewed by means of cuttings that will take a few weeks to root in the propagator.

GARDENIA JASMINOIDES (syn. *G. florida*, *G. radicans*). This is the name for various forms of gardenias that were at one time considered to be distinct. They vary in height from 2 ft. upwards, and have either double or single flowers, growing individually at the ends of branches. They invariably possess that intense fragrance associated with this genus. Leaves are generally bright and green.

G. J. FORTUNIANA. This is just one of several varieties that has somewhat larger flowers than the species.

Ixora

These rubiads are wonderful stove plants, excelling the qualities of the renowned gardenias both in appearance and show of flowers. They belong to a group that bears heads of blossom much like the viburnums of the open garden. They need much warmth and lots of moisture during summer, considerable warmth and some moisture in winter. They root easily from cuttings in early spring, and should be potted-on as needed in John Innes compost, to which is added moss peat. They need fertilisers at maturity. Flowers begin to arrive on quite small plants, usually starting in May or June. Older plants need repotting once every two years. They generally stay in good shape, but can be cut back after flowering should it be necessary.

IXORA CHINENSIS. Grows about 3 ft. high. Comes from China and has nice handsome leaves. Flowers are orange and bunched. There are pink, yellow and white forms among the garden cultivars.

I. COCCINEA. This is a beautiful species from the East Indies, which bears large heads of scarlet flowers above heart-shaped foliage.

I. C. FRASERI. Has lovely large leaves and deep salmon flowers. It grows a bit more freely than some. Is but one of several cultivars.

I. FULGENS. The leaves resemble those of the willow. It used to be grafted on to another species because of its reluctance to grow. Height is about 3 ft. Flowers are orange-scarlet in large bunches, but rather thinner than in some species.

I. WILLIAMSII. This is mentioned because it is the most likely one to thrive in the warm greenhouse rather than the stove. Flowers are large bunches of orange-red shades, while foliage is dark green. There are many hybrids and cultivars.

* * * *

BURCHELLIA CAPENSIS. Has a few deep scarlet flowers, in a bunch, in spring. Will thrive in the warm greenhouse, and eventually attain 5 ft.

COFFEA ARABICA. Arabian coffee. One species of a genus that yields the coffee beans. Makes an interesting and quite nice small tree when grown in the stove. Generally raised from seed. Evergreen leaves are shining green and white fragrant flowers are borne during autumn.

HOFFMANIA GHIESBREGHTII. Velvety green foliage with purple undersides. Attains 4 ft. Yellow flowers are spotted red.

H. REFULGENS. Grows about 2 ft. high and has smallish green leaves tinged red. Flowers are pale red. Will grow in warm greenhouse.

PAVETTA CAFFRA. This is closely related to the ixoras, and needs the warm greenhouse rather than the stove. Flowers are white, and freely produced during late summer. Foliage is bright and attractive.

P. INDICA. An evergreen shrub of good appearance that bears fragrant white flowers in smallish bunches. Needs stove treatment rather than warm greenhouse. Grows about 3 ft. high.

PENTAS LANCEOLATA (syn. *P. carnea*). A soft-wooded shrub of smallish habit, which bears bunches of pale purple flowers after the manner of bouvardias, in late summer. Can be pinched back almost at random, and young shoots root easily in the propagator. Should be given fertilisers once pots are filled with roots.

P. L. KERMESINA. Has bright pink flowers and grows about 2 ft. tall. Leaves are variable in size and of an attractive soft green.

PORTLANDIA COCCINEA. Has very handsome, perpetual foliage and impressive, individual, largish, red flowers with yellow anthers. Height is not more than 2 ft. and season is summer. Needs the warmest part of the stove, and the treatment described for ixoras.

P. GRANDIFLORA. This grows into a tall shrub and produces beautiful foliage, together with most impressive white flowers. Needs much warmth and moisture. Like most flowers that are redolent during dark, they emit a powerful fragrance that is a bit too much for sensitive nostrils.

POSOQUERIA FRAGRANTISSIMA. Has shining green leaves with yellow veins, and bunches of white flowers somewhat after the manner of stephanotis, with tubes to the individual flowers like the tobacco plant. This and the other desirable species appreciate the stove. A peaty compost is better than a loamy one. They can be raised from cuttings in spring and potted on when rooted. They must have light or they grow straggly. Tops are pinched out when growths are a few inches long. They flower in summer. Chief thing is to cut back after flowering, so that new shoots are formed before the comparative resting period of winter. They improve with age.

P. LONGIFLORA. Grows to 5 ft. Has very long-tubed, trumpet flowers that are scented. Several flowers arrive on each bunch during summer. Treatment as for above.

P. MULTIFLORA. Has lovely soft leaves and white flowers not quite so long but more spready than the above. (There are other species most desirable if the stove conditions can be provided.)

PSYCHOTRIA JASMINIFLORA (syn. *Gloneria jasminiflora*). The wild coffee is but one of several first-class evergreen shrubs that have the leathery leaves and bunches of white flowers somewhat after the manner of jasmine. Needs the usual treatment for the family, but will more likely thrive in the warm greenhouse than in the stove.

RANDIA MACULATA (syn. *Gardenia stanleyana*). This is a species that was once listed as a gardenia, but has now been segregated. It is a remarkable shrub that has flat branches and flowers, almost like small tulips with reflexed petals. They are white with purple spots.

RONDELETIA ODORATA (syn. *R. speciosa*). Is but one representative of yet another genus of this most attractive family. The orange-red flowers arrive in terminal bunches, rather like the viburnums, and are fragrant; they last a long time, cut or on the plants. Warm rather than stove conditions suit them best: they then grow more sturdy, and slowly develop through the years. Leaves are typically leathery and pleasing.

CHAPTER 22

Verbenas (Verbenaceae)

THERE are many species of verbena that only need the protection of a cool greenhouse, and none really worthy of a place in the stove or warm greenhouse. However, the family does include some genera that are most desirable, and in particular the clerodendron. Some species are climbers, some are shrubs or trees, some are hardy, some need some protection and several are of tropical origin: I merely mention *Cornutia*, *Duranta*, *Lantana*, *Premna* and *Vitex*. Ironically, *Premna integrifolia* is commonly known as the headache tree, probably on account of the scent it emits, but it poses a problem as to whether to include it in this selection, for there are so many other subjects more worthy, but which have to be omitted.

Clerodendrum

CLERODENDRUM FRAGRANS PLENIFLORUM. This is a weaker plant than the *C. speciosissimum* described below, and is deciduous, but it is often preferred because the pinkish white flowers are exceedingly fragrant, somewhat like clove carnations, and are double.

C. SPECIOSISSIMUM (syn. *Clerodendrum fallax*). A free-growing, evergreen shrub that gives much pleasure wherever grown. It has fresh-looking leaves up to a foot long and almost as wide. Above these, from the top of the shoot, come upright panicles of bright scarlet flowers, each about 2 in. across, the whole looking rather like a refined rhododendron. This species, as with practically all verbenas, can easily be raised from seed sown in autumn or, if one prefers, cuttings may be rooted about midsummer, having been taken off with a heel. Flowering season is early autumn, and normal height 3 ft. John Innes compost suits them well. Give plenty of water in summer, but in winter let them get almost dry. After flowering, prune the shoots hard back, as for bush roses. Repot in February. These shrubs have the ability to bear flowers when young or old, and can easily be kept to size.

C. SPLENDENS. An evergreen shrub that will sometimes produce shoots, to be pinched back or trained to a support. Usual height is about 4 ft. It produces heart-shaped leaves about 4 in. long, and bears large bunches of scarlet flowers during summer. Not an easy shrub to grow.

C. THOMSONAE BALFOURII. (syn. *C. magnificum*). A vigorous climber with heart-shaped leaves 5 in. long. The flowers are mainly scarlet and arrive in panicles. This was common in large gardens at one time. Strikes well from cuttings, taken with heels when about 6 in. long during March. Keep watered and syringed until September and then withdraw moisture until there is flagging. Keep fairly dry through winter. Prune straggly shoots back in February, and then start to syringe to obtain new growth. Other good species are *C. incisum*, *C. japonicum* and *C. ugandense*.

* * * *

PETREA RACEMOSA. The purple wreath and *P. volubilis* are sometimes confused, although they are quite distinct. *P. racemosa* is a low climber, and can be grown much like *Asparagus sprengeri*. The leaves are shaped like those of the plum. Flowers of lovely blue shades come in racemes several inches long during summer. Heat and moisture in plenty during summer are the chief needs. Can be propagated by new shoot cuttings in spring.

CHAPTER 23

Miscellaneous Plants

IN this category are a number of plants that do not justify chapters unto themselves. Some are insignificant. Others are very beautiful or interesting, or both, and quite worthy of a place in a mixed collection.

In most cases there is only one representative of each family, but there are a few families which have plants worth mentioning from several genera: for instance, in the pea family (*Leguminosae*), most of which are completely hardy, though the more exotic members will thrive in a cold or cool greenhouse. A typical example of the latter is wattle, sometimes known as mimosa or acacia. However, there are a few tropical representatives needing the stove or warm greenhouse. The main ones worth growing are mentioned in the following list.

Another family with a few, but good, plants in it is the *Marantaceae*, closely related to the cannas, but still distinct from them. The marantas are useful stove plants and are often recommended for living rooms. They withstand more shade than many of the other stove occupants. Compost should have a preponderance of peat, although some loam in the mixture is recommended. They like plenty of water from spring until autumn, less of it from autumn until the end of the year, and almost dry from then until February. They should be repotted at this time and nursed into growth, and then syringed freely. They are increased by rhizomes or tubers at potting time. Calatheas and ctenanthes need similar treatment. Other genera in the family are *Monotagma*, *Phrynium*, *Saranthe*, *Stachyphrynium* and *Stromanthe*.

As there is an interest in the banana family (*Musaceae*), which for the most part needs much heat and sunshine, they are also included, but are not considered to be ideal for stove or warm greenhouses unless there is a lot of room in which they can grow. However, the paradise flowers are in the family and these simply must be included. For the fruiting bananas, a bed or a very large tub is necessary. John Innes compost is suitable for them. They are propagated by suckers taken from the old plants whenever available, at any time of the year. There are hardier, non-fruiting, banana varieties, some of which will grow in the mildest parts of Britain.

Finally, there is the family from which ginger comes, the *Zingiberaceae*; whether one wishes to grow members of this extraordinary family of tropical plants or not is a matter of taste. They are quite distinct in character to all else, although they have semblances to other subjects in various ways. The roots are rhizomatous, while the individual flowers are somewhat like cannas, but in many instances they appear in a spike like a pineapple, A hardy representative in the rock garden is the genus *Roscoea*. Like plants in herbaceous borders, the tops perish each season. Some need the stove throughout, while others flourish in a warm house. Large pots or receptacles are their permanent homes, and new plants are obtained by carefully dividing the rhizomes in early spring. Ordinary compost is suitable, the John Innes potting mixture being ideal. Older plants are either repotted, or put into slightly larger ones. They are given much moisture when growing fast and are kept moist when flower stems and foliage fade. These are chopped off in late winter. If this is carried out too early, the roots will suffer.

ADENANTHERA PAVONINA (*Leguminosae*). The peacock flower fence is an evergreen shrub that will attain 5 ft., having divided leaves and white and yellow flowers in a mixture during summer. The resulting seeds are often used as necklaces. Usual warm greenhouse treatment will suit it; it can be propagated from cuttings, or from seeds, which should be chipped or soaked first.

ARDISIA CRISPA (syn. *A. crenulata*) (*Myrsinaceae*). The spear flower is a neat evergreen that likes the warm house, and a good representative of many species. The flowers are insignificant, but are fragrant, and first arrive in large bunches in June followed by brilliant red berries comparable with haws. This shrub needs practically nothing more than ordinary treatment—not even pruning.

ARRABIDAEA MAGNIFICA (syn. *Bignonia magnifica*) (*Bignoniaceae*). A climbing shrub with panicles of lovely, blush crimson flowers that have light yellow throats in summer. Usual warm greenhouse treatment (including propagation) suits it well. It is evergreen, and should therefore never be allowed to dry out completely in the resting period. The shoots should be thinned when growing in order to get quality flowers, and also need to be pruned back after the summer season. The family contains other genera more suited to the cool greenhouse, some erroneously named bigonias. The family includes the evergreen *Jacaranda mimosaefolia*, with the lovely ferny leaves and striking blue flowers. These start to arrive when the tree is about 12 ft. high.

BOUGAINVILLEA GLABRA (*Nyctaginaceae*). Is one of two main species, the other being *B. spectabilis*; each producing a few varieties, differing in colour shading and growing ability, but all liking plenty of root room. If planted in a border and trained up a trellis against the wall, or behind the staging, the main stems can be trained out and each year, after being pruned back in February, new long shoots will form, being studded overall with small flowers that are covered by green bracts, which turn to a startling metallic rosy-mauve shade and last for many weeks in summer. They will grow in large pots but do not yield bracts so freely.

BRACHYSEMA LATIFOLIUM (*Leguminosae*). An evergreen climber to grow against a post which should be planted in a well-drained bed in the warm greenhouse, or in a tub. Needs judicious pruning to keep in shape. Flowers are crimson-scarlet and arrive in spring.

CALATHEA ORNATA (*Marantaceae*). The varieties of this species are the most popular plants, often listed or sold under the name *Maranta*. They are variable in height and character, but most often have green leaves decorated white, with purple undersides. Flowers are white and violet, of much less importance than the leaves.

C. O. ALBO-LINEATA. Has pairs of silvery stripes in the oval leaves. Ideal for baskets.

C. ZEBRINA. The leaves look like refined hostas, but are of apple green, and marked a beautiful brown between the veins.

There are many more species from Brazil and other tropical places.

CASSIA FISTULA (*Leguminosae*). Has multiple leaves a foot long and bears hanging racemes of fragrant golden yellow flowers, giving it one of its common names, golden shower; others are the pudding pipe tree or senna. Later, the bean-like pods develop. It is an evergreen tree and needs a well-drained border or tub with usual soft mixture. There are several other good species and varieties.

CISSUS DISCOLOR (*Vitaceae*). This is a very free-growing climber that is useful for being trained up pillars in the stove, or on a trellis against a wall. The chief attraction is the wonderfully variegated, elongated, heart-shaped leaves that are deeply ribbed. The base colour is deep shining green, marbled with silvery overtones, and there is a pinkish red hue while the leaves are young. They will grow in well-drained compost of any sort, and need large pots when mature. Once roots are well-formed they must be fed regularly. Plenty of moisture in summer, and less in winter, is the usual treatment. If exposed too much to sunshine the variegations fade. It can be raised from cuttings of young wood growths, put singly into small pots and plunged in the propagator in spring, when they will readily root. After potting into a 6 in. size, insert sticks for supports until they grow tall enough to be attached to the main supports. It should be lightly pruned into shape in March. There are other species of some variation that will grow in the stove or warm greenhouse, including *C. javalensis* (now *Vitis javalensis*) and *C. gongylodes*. A close relative is *Leea amabilis*, which is virtually a shrub, with feathery leaves that are striped white. Height 3 ft.

CLITORIA TERNATEA (*Leguminosae*). The butterfly pea is an evergreen climber up to 12 ft. with pinnate leaves, and bears bright blue flowers during summer. Needs a large pot or tub with usual compost. It will thrive in a warm greenhouse. Easily propagated from cuttings or by seed.

COLEUS THYRSOIDEUS (*Labiatae*). Whereas the popular garden and greenhouse coleus is grown for the variagated leaves, the winter-flowering coleus is mainly cultivated for the scintillating blue panicles of flowers that sprout from small branches during winter. But the leaves are also attractive and make a good plant up to 2 ft. high. The stove is the best place for this species. Keep moderately moist in both winter and summer. Cut back in spring, and root cuttings of the new shoots. Pot on as usual in the John Innes mixture as necessary. The more popular strains of various hybrids and cultivars, grown chiefly for the marvellous leaf colours during summer, are mostly raised from seed sown in spring. But they can also be propagated from cuttings of young shoots taken any time and rooted in light compost. These are less tender than the winter-flowering species. They like a lot of water in summer and very little in winter. If any grow straggly, or too tall, the tips should be pinched out to cause bushiness.

CTENANTHE LUBBERSIANA (*Marantaceae*). Only grows about half as high as the following, and leaves likewise are half the size. They are beautifully marked yellow on a deep green background, with a much paler shade of green underneath.

C. OPPENHEIMIANA (syn. *Stromanthe porteana*, *Calathea oppenheimiana*). Has beautifully patterned leaves of laurel shape, on thin wiry stems, typical of the maranta family. They are often over 1 ft. long, while the mature plant attains 3 ft.

DELONIX REGIA (syn. *Poinciana regia*) (*Leguminosae*). The peacock or flamboyant flower is a deciduous, umbrella-topped tree up to 15 ft. tall. The leaflets form a herring-bone. Scarlet flowers arrive in April and pods of a woody nature develop later. It needs a stove and the usual compost with sufficient pot room.

DOMBEYA VISCOSA (*Sterculiaceae*). An evergreen tree that grows to 15 ft., but can be kept dwarf by restriction of room and pruning. It has heart-shaped leaves with serrated edges and bears heads of fragrant flowers, of pure white with a crimson base, in summer. Pot and prune in March. Needs ordinary compost and usual greenhouse treatment, i.e. plenty of water in summer, less in winter. This is a typical representative of a smallish family of tropical and temperate trees, chief of which is *Theobroma cacao*, which yields cocoa. *Fremontia* is a garden shrub, while among the sterculias are other stove trees that need too much room to warrant mention.

EUGENIA MALACCENSIS (*Myrtaceae*). The Malay apple is

closely related to the lovely myrtles, and is one of a few species that needs stove temperatures. It is evergreen, grows up to 10 ft. high, and bears scarlet flowers in June followed by edible fruits. Another species, *E. aromatica*, yields the cloves of the kitchen, but is not a very picturesque tree. It should be grown in ordinary compost in pots according to the size of small trees. Keep moist in winter. Give water and syringe freely in summer. Generally propagated from cuttings.

HEDYCHIUM CORONARIUM (*Zingiberaceae*). The fragrant garland flower grows to 4 ft. and bears spikes of fragrant white flowers in late spring.

H. GARDNERIANUM (*Zingiberaceae*). The ginger lily attains 6 ft. and will grow also in a cool greenhouse. The flowers are yellow, appearing in summer.

H. GREENEI. Grows to 5 ft. It has long leaves and shortish spikes of red flowers clothed with bracts. This one will flourish in a warm greenhouse. All need some shade.

HELICONIA BIHAI TRIUMPHANS (*Musaceae*). The false plantain is but one of several species and varieties of plants that produce lovely foliage. This representative has leaves up to a yard long, striped black on a dark green surface. Bracts are borne as well as flowers. The height is about 3 ft., and it needs much the same treatment as bananas.

HIBISCUS ROSA-SINENSIS (*Malvaceae*). The blacking plant is a more attractive and refined plant than the shrubs of the open garden. The flowers are like large petunias, often 5 in. across. The main colour is rose red, but there are also white, as well as double-flowered, variations. Main season is summer. It likes very warm and moist conditions. and should never be allowed to get completely dry at the roots. Propagation is by half-ripe shoots, that form after pruning at the end of the flowering season. John Innes compost is suitable. It likes shade from hot sun. Old plants are the most floriferous.

H. COOPERI. The foliage of this variety is an improvement, especially when variegated.

Other genera in this family (the *Malvaceae*) that need greenhouse protection are *Abutilon*, *Goethea*, *Malvaviscus* and *Pavonia*.

MALPIGHI GLABRA (*Malpighiaceae*). A stove evergreen that attains small tree size and then bears cherry-like fruits; which give it its name of the Barbados cherry. Its leaves are bright and flowers are rosy purple, appearing from March to September. There are other species in the same genus, which are hardly exciting enough for one to be anxious to obtain them.

MARANTA BICOLOR (*Marantaceae*). Remindful of the elliptic shape of the smaller peacock feathers. They are about 5 in. long and shiny, and have clear quilled markings of light green on a dark background, with a purple splash on the under-sides. Flowers are white.

M. ARUNDINACEA. It has racemes of white flowers, while the leaves are about 6 in. long, of no exeptional quality. It is the source of West Indian arrowroot.

(Several plants named *Maranta* are now in the genus *Calathea*.)

MIMOSA PUDICA (*Leguminosae*). Known as the humble plant, the well-known annual whose leaves droop when touched, and is distinct from acacia. It grows well in the warm greenhouse without any special needs.

M. SENSITIVA (*Leguminosae*). This is the sensitive plant which responds much the same as the foregoing, but climbs up to 4 ft. It has purple flowers. More interesting but less sensitive than the humble plant.

MUSA CAVENDISHII (*Musaceae*). This grows about 5 ft. high, and bears seedless fruit. The leaves are glaucous and about 2 ft. long, often a foot across. The chief needs are enough room, good compost, warmth and moisture.

M. PARADISIACA. This is the species that bears plantain bananas and grows up to 5 ft. The leaves are often several feet long and half a yard across. There are many cultivars. It needs a roomy, tropical home.

NEPENTHES COCCINEA (*Nepenthaceae*). One of the easier species of pitcher plants to grow in the stove house. Although there are many hybrids and cultivars, they are rather difficult to cultivate, especially to a state of perfection when they will produce the pitchers that give them the common name. Those appendages at the ends of very long leaves, that are somewhat like those of aspidistras, form and hold water after the manner of hot-water jugs, with lids on top. The lids stay open, except when insects crawl in the top, when they close, trapping them until they fall into the water. There the insects decay and their bodies are absorbed into the systems of the plants. The shapes of the pitchers and their colourings vary with the various kinds. But the chief need of all is plenty of moisture, even in winter, which makes them difficult to grow in a general collection. Another problem is that the roots are poor and look so dead that they may be mistakenly removed, causing distress. Some are epiphytes. The hints upon the cultivation of the bromeliads are helpful regarding mixtures and potting, but not the watering. These plants need the roots watered almost all the time. They also like a lot of light. They will ramble, and shoots will attain varying lengths up to 5 ft. Old gardeners shortened these a few times in order to build up the structure so that the pitchers were obtained. They are propagated by cuttings of one-year-old shoots put in the warm propagator in spring.

PANDANUS VEITCHII (*Pandaceae*). The screw pine is one of several varied species, so-called because of the leaves that grow from the main stems in screw-like fashion. This representative of the family makes a nice plant of modest proportions, but will attain large size if given the opportunity. Leaves are up to 2 ft. long and beautifully variegated with white markings. They are tough and have spiny edges. The plants thrive in John Innes compost, and often work their way up out of the pots. When they become shabby it is best to cut them down, when suckers will grow out of the roots, which can then be severed and given normal stove house treatment.

PILEA MICROPHYLLA (*Urticaceae*). A member of the nettle family almost as easy to grow, if it gets the warmth of the heated greenhouse. Will thrive in any soil, and can be propagated from cuttings taken off the ends of shoots. The common names of artillery plant or pistol plant are given because the unexpanded buds, if moistened or disturbed, will explode and give off clouds of pollen. A better plant is *P. cadierei*, with oval leaves that are painted silver. It is often referred to as the aluminium plant. There are varieties, a good one being 'Moonglow', whose leaves are black and green, giving a fuzzy carpet effect. All are squat and tidy.

PLUMBAGO ROSEA (*Plumbaginaceae*). A handsome evergreen pot plant that grows about 2 ft. high, and bears in winter heads of rosy-scarlet flowers that look somewhat like phlox. To grow well, take cuttings of young shoots in March, after a brief rest from flowering, and root in the propagator. Then pot in John Innes mixture, and when three or four joints have formed, pinch out the tips to cause branching. Keep them in the stove and make moist by syringing. If they grow spindly, pinch out the tips once again. Always keep near the glass, but shade during hot sunshine. Give fertiliser when the plants are growing freely, rather than increase the size of pots. While we are often told to use it for trellises, it is best regarded purely as a pot plant, and is quite distinct from the blue *P. capensis*. This one originally came from the East Indies.

RAVENALA MADAGASCARENSIS (*Musaceae*). The traveller's tree takes up far too much room in a greenhouse, and needs lots of moisture. The chief attraction is the enormous leaves which are often 10 ft. long.

SELAGINELLA PALLESCENS (syn. *S. cuspidata*, *S. emmeliana*) (*Selaginellaceae*). A stove plant that resembles a tiny fern, growing up to 6 in. high. The branched stems are covered with tiny, bright leaves. It should be given a mixture of peat and sphagnum, rather than loamy compost. It loves shade, and needs lots of water in summer and moderate amounts in winter. Can be propagated by cuttings taken in spring and summer, or by division.

The species *S. kraussiana* will flourish in the warm greenhouse, and forms what looks like a ball of moss. There is also the resurrection plant (*S. lepidophylla*) sometimes sold as a novelty. The plant looks like a dried-out ball, but when placed in water, the fronds uncurl into greenness.

SOLANUM PENSILE (syn. *S. amethystinum*) (*Solanaceae*). This is one of a few of the potato and tomato family that need a warm greenhouse. It bears deep blue flowers with white centres in May, followed by cherry-like, pale blue fruits.

An orchid house.

Greenhouse showing chain lath blinds mounted on galvanised iron runners

▲ Cedarwood conservatory.

▼ Small greenhouse/conservatory.

A border containing ordinary soil is most suitable. A comparative resting period is necessary in winter, and the pruning of straggly shoots is best carried out after fruiting. Another species needing similar treatment is *S. wendlandii* that bears lilac flowers during August.

There are many other species and varieties of ornamental solanums, including the winter cherry, *S. capsicastrum*, but only suitable for the cool greenhouse.

STRELITZIA REGINAE (*Musaceae*). The bird of paradise flower, or bird's-tongue flower is, surprisingly, a member of the banana family, most popular among florists on account of the pouch at the end of a stiff stem, from which emerges alternating petals of blue and orange, giving the flower its common name. The height of a good plant is 4 ft. and it will bear several leaves suggestive of paddles with blades over a foot long, above stiff stems, all of an apple green shade. Coming from South Africa, it needs less heat than bananas, the warm house suiting it perfectly. Good rich soil, with plenty of water in summer and a limited amount in winter, are its chief requirements. Propagation is chiefly by division, but I have raised them from seed. There are other species, as well as several varieties of this one.

TAPEINOCHILUS ANANASSAE (*Zingiberaceae*). As the species name denotes, this plant resembles the pineapple to a considerable extent. It grows several feet high and sends up a flower stalk about 3 ft. tall, surrounded by rich red bracts and inset yellow flowers. The stove is necessary, as it needs heat and moisture while growing.

VELLOZIA CANDIDA (*Velloziaceae*). A representative of a small family of plants of singular character. This one comes from Brazil and needs stove protection. The leaves are tufted and not unlike in shape to the familiar day lilies, but are tougher and inclined to be prickly. The flowers appear individually in summer on long thin stalks, each with six petals forming a flattish 'dish' of pure white. The chief characteristic of the plants is that the old leaves form a mat at the base that retains moisture when syringed. This, rather than the roots, gives them sustenance, in much the same way as the cups water the bromeliads. In other respects their needs are similar.

ZINGIBER OFFICINALE (*Zingiberaceae*). The ginger plant, growing to about 3 ft. high. Flowers are yellowish, with purple spots, accompanied by pale greenish bracts. The rhizomes are the parts grown for the ginger used in commerce. Needs the stove and treatment described above.

Part Three

Please note that neither by inclusion, nor omission, must the name of any manufacturer, factor, firm or person mentioned in the following pages be regarded as a recommendation, or otherwise. It is possible that there are national and local suppliers who can offer such products, plants or seeds at more competitive prices and give better service. The addresses are primarily given for the benefit of those who do not know where to find their greenhouse requirements. All literature should be studied, and the reader is strongly advised to compare qualities and prices.

CHAPTER 24

Notes upon Greenhouses and Conservatories

IN the first chapter I gave a brief description of what I regard as being the ideal small greenhouse. But I do realise that it may differ from the vast majority of the infinite variety, which ranges from the gigantic glass palaces of the commercial men, down to the tiny, shoddy affairs that so often find their eventual vocation as homes for deck-chairs and spiders.

It would be remiss of me to praise or condemn the various shapes and sizes of many manufacturers out of hand. The best way to approach the subject in the most practical way, I think, is to consider the chief features, so that when buying new structures or converting old ones into suitable homes for tropical plants, we do so with a degree of forethought.

Site

The site and the position where the greenhouse has to go will be a determining factor regarding shape and size. It should be in a sunny position: not only does that save fuel costs but also, although we shade the plants in summer, they benefit from all the winter light they can get.

A lean-to, or three-quarter-span house, upon the south side of a wall, is probably the most economical and satisfying type. If it is an integral part of the residence, it can be a conservatory, or a real plant room. The difference between them is that a conservatory is built for the comfort of owners, the plants being merely part of the decorations, as it were, while the real plant house, on the other hand, is not so comfortable for those who bask, as for the vegetation that inhabits it. For example, the pathways are more extensive and more pleasingly decorative in the conservatory than they are in the greenhouse. In the latter case they are often very narrow, and the whole floor space is such that moisture is retained to create humidity. Plants are happier in this atmosphere.

If the plant house is a unit apart a span-roofed building is undoubtedly preferable. It should not be farther from the residence than is absolutely necessary and, if possible, should be given a warm and level sunny site. It can be a thing of beauty and become a wonderful asset.

Contrary to a common belief, the span-roofed greenhouse should have the ridge running from east to west, rather than north to south. There is no doubt whatever that where one long side faces south it is much warmer than when the long sides face east and west. This has been proven again and again.

Walls

For heated greenhouses, the lower part of the upright walls should always be solid, preferably being of bricks, forming a width of not less than 9 in. Those with glass panes right down to the ground let in too much cold in winter, and get too hot too soon in summer. They also look wrong. The types with sloping sides, generally made of Dutch Lights, are the worst of all. While they may be excellent, and economical, for growing market crops of lettuces and tomatoes, they are practically impossible for our purpose. On the other hand, those with brick walls about 30 in. high not only save fuel, they keep the main structure of the greenhouse in good shape, and also look substantial and nice. Should the bricks of the greenhouse match the bricks of the residence, they look as if they really belong, and become most attractive.

If the plant house is to adjoin the residence, in lean-to style, it pays, in the long run, to have it made to order, rather than obtain a standardised product. There are still firms that will do this, in spite of the thousands of other greenhouses that are churned out like cheap cars.

In fact, I feel sure that getting a good greenhouse is by far a better investment than buying a car of any quality. For the amount you must pay for a moderately priced vehicle, a marvellous, almost luxurious, conservatory or greenhouse can be obtained. Within a few years, the value of the car will have dissipated, whereas the greenhouse will have increased the value of the property by an amount far exceeding the original outlay.

However, if it has to be a span-roofed house, there are good standardised buildings available. Alas, there are some shoddy ones too. This sort of investment needs more careful consideration than the specially-made product, for the latter is almost invariably excellent, incorporating all the features the manufacturer takes a pride in providing.

While there is no doubt that a large greenhouse is easier to manage than a smaller one, from the viewpoint of maintaining an equable temperature, it is not advantageous to have a tall structure. The roof need not be high-pitched. As long as there is no run-back of rain under the overlapping panes of glass to create drip, it can be a gentle slope. High-pitched roofs are good where there is little fire heat provided, but for a stove or warm greenhouse, the heat collected at the apex is wasted. Another reason for high-pitched roofs is that snow can weight down the structure of a low-pitched one. But if we keep it heated, that contingency does not arise.

The panes of glass should be generous in size. Large, commercial houses often have them a yard long and 2 ft. wide. For a moderately sized structure I regard those that are approximately 16 in. wide and 18 in. long to be quite good. They enable a considerable amount of light to enter, and yet are small enough to need sufficient framework to retain a solid structure.

Framework

There are, as everyone knows, frameworks made of many materials. At one time the most usual were of timber. The modern commercial houses are mostly of aluminium or aluminium alloys. There are also concrete structures as well as others of iron.

Timber. The commonest now in the small house ranges are made of 'red cedar'. I understand that this is actually *Thuja plicata.* If properly matured, it remains sound for many years without being painted. Due to the non-upkeep quality, lightness and straight grain, it is the timber that is most boosted.

However, Baltic redwood is quite often the alternative. This is derived from Scots pine and is really strong, even lacking the common fault of splitting when nails are driven into it. It is less costly, too. But it has to be treated with preservatives very frequently, or rot will surely develop. While there is no doubt that western red cedar is more foolproof, even that is improved in appearance when painted.

Some of the flimsy greenhouses not only lack strength because of weak rafters, but sometimes the sash-bars are made of immature wood. The surest way of obtaining a good, long-lasting building is to patronise a firm with a good reputation for first-class workmanship.

Aluminium. An increasing number of new small greenhouses are being made with aluminium frames. Some look very smart. They generally cost somewhat more than the wooden ones, but once installed will last for ever, without much attention. It is as well to study carefully whether these are the right shape and size, before we make a definite decision. We must remember that it is needed for rather important plants, and that they should never be cramped where they need most room—at head level.

The glass above the brick wall base should go straight up to the eaves for quite a distance—say, 3 ft. above staging level. We must envisage the needs of our plants, rather than indulge in more attractive, though less practical, designs.

The next point to look for is the ease with which a broken pane of glass can be replaced. This may not occur often, but it can be annoying if you have to strip the house down, or obtain elaborate steps in order to overcome this problem.

Still another problem I have found, is the fixing of wires for climbing plants. Some firms make provision for this, by making suitable holes in the aluminium frames. However, there are others in which it would be dangerous to try to fix wires to the bars. Aluminium will not warp, like wood, but it will bend under stress, and it is often fragile.

Quite often iron is used to strengthen the main aluminium structure. Then there are greenhouses in which wood, and some sort of metal, are combined. Regarded generally, they are good, provided they are genuinely explained, for the manufacturer has obviously taken considerable interest in the technicalities involved, rather than using them as mere selling points.

I am a gardener, and do not profess to know the theoretical aspects of greenhouse construction, any more than I do those of motor-cars, but I do look at the prospect with critical consideration, regarding longevity, comfort, size, cost and suitability for the inmates, before making a decision.

Ventilation

All good greenhouses have ample ventilators that work. The best ones have them along the whole length of the roof ridge. They most often have them throughout the length of the sides as well, usually of a pivoting type, so that all the glass between the solid wall and the eaves can be adjusted to let in fresh air when it is necessary.

In the chapter on orchids, I stated that if box ventilators were made in the brickwork, the others were unnecessary, and when we converted the structure to an orchid house, we closed them for good. But, if we are mainly concerned with the growing of plants other than orchids, we do use the ventilators, as detailed in the second chapter. When we grow a few orchids among the other plants, these have to put up with what the plants want. A few orchids are more likely to flourish among plants, than plants are in an orchid house, although there are many that will do so.

The need for greenhouse ventilators is to control inflow and outflow of air to regulate temperature, rather than provide oxygen, carbon dioxide and all the rest of it, for the plants' welfare. Plants will grow in closed bottle gardens, therefore they will surely grow in greenhouses where it is almost an impossibility to stop an exchange of air between the outside and the interior.

But the chief thing is to avoid draughts. For some reason, people, and plants, cannot stand draughts of cold air rushing into warmth. Therefore, if the box ventilator is the only opening during winter, there is a minimum of cold air induction at a place where the plants are not near, and where it is tempered by its passing over the hot-water pipes. In cold weather, I would hesitate to open the box ventilators and the ridge ventilators at the same time. That would create some draught. But when the outside temperature is near to the one I want indoors, then it would not hurt at all. What I am trying to explain is that intelligence should be employed over the provision and the use of ventilators. Think of draughts. Think of escaping heat when it is cold outside. Think of bringing the inside temperature down without draughts at certain times of the year, and at others, when it is very warm outside, opening up freely without losing the humidity. It all depends upon the time of year and how the winds blow.

Staging

Those who have read the previous observations on staging will gather that I favour the wooden slatted type, rather than a solid shelf. My main reason is, apart from the fact that air circulates more freely, that the moisture from the ground gives the plants buoyant humidity. Furthermore, we can also grow many plants under the staging, if the soil of the floor is suitable. Ferns, mosses, and some plants that love shade, do not mind obtaining their moisture and nourishment from the spillings given to the plants over their heads. And the very fact that they breathe has a beneficial effect upon the ones above.

However, there is a difference between good wooden staging and that often put in cheap greenhouses for economy's sake. Just a few slats placed across the side of the house, held up by the ends being fixed to the cheap greenhouse wall with a leg in the middle to stop the hammock effect, are ill devised. Plants in pots are heavy. The staging should be properly constructed with a strong framework and be held up by good, solid pillars or posts. If the lower walls are of brick, they can be used to support the staging where practical, but there should always be a gap between the wall itself and the staging, except where the struts are placed.

But more good greenhouses are fitted with solid benches or staging than are with open wooden ones. I have worked in ranges where these exist, and have grown just as good plants upon a bed of ashes or shingle as upon the open-worked ones. The chief drawback is the impossibility of having them growing in the soil as well, for solid staging completely cuts out the overhead light. And there are watering difficulties.

It matters not whether the floor of the staging is of slate, or concrete, or galvanised iron, as long as there is something less impervious spread over the top. What matters more is that the structure is solid enough to hold up the weight of the glass house structure independently. Struts can be placed on the brickwork, but never upon a more flimsy base, such as wood.

As shown in Fig. 1, if there is an interior stage in addition to the one upon the front side of the pathway, its edge is best not on the same level as that of the outside one. In the case of the tiered shelves, the front one is lower. But if the interior stage is to be a solid table, then it should be a little higher than the front one. It then raises the plants desirably nearer to the ridge, so giving them more light and air, and warmth. The only drawback to this is the too obvious space below the staging, if it is set very high (see Fig. 2).

You can, of course, create an undulating effect upon the middle staging by putting on top of it another framework, but the usual way to do that is to place some of the pot plants on other pots turned upside down.

Water

In the olden days water had to be conserved, so all the rain was collected by gutters and pipes leading to an underground cistern, from which it was pumped into other, smaller tanks inside each greenhouse. That has gone out of fashion. Lack of labour, the ease with which mains water can be laid on, and an insistence that it was bad for plants has influenced this.

The belief was that water needed to be of the same temperature as the greenhouse to suit the plants. Now we have been told that experiments have proved that plants grow equally well when watered straight from the tap. Tap water is not exactly chilling. But if used a good deal it can reduce the temperature of the house, and more artificial heat is necessary to keep that up.

I believe that soft water is preferable for stove and warm greenhouse plants, when it is of the same temperature as the house. It is best obtained by fitting proper guttering so that it flows into a sunken tank within the house, and so arranged that there is a safe overflow to a drain outside. It should also be planned so that the same tank can be refilled by merely turning a tap leading from the mains.

Regarding the argument that the regular use of collected rain water in an inside tank breeds diseases, I agree that it might if the underground cistern is never cleaned, which is difficult, nor the surface one in the greenhouse swabbed occasionally, which is easy; this form of water conservation might then contaminate plants. But I also hold the theory that if water can conduct diseases it can also spread bacterial activity, which is conducive to plant life. Strict sterilisation for plants is a bad thing. Their home is in soil or, in the case of our tropical subjects, often in accumulated muck. They are themselves natural sweeteners.

Blinds

As shading of some sort is absolutely necessary, provision should be made at the outset, and there is still nothing better than blinds that are fitted to the roof in such a way that they can be raised and lowered easily with cords when desired.

When it cannot be done manually, the usual treatment is to dress with Summer Cloud or some such branded liquid or powder, watered down to a recommended thinness and applied with a whitewash brush or sprayer. It is not as good for plants, since they cannot be exposed to full light on dull and sunless days.

There are also available automatic blinds that are rolled up, or let down, by an electric motor controlled by a photo-electric cell. These are wonderful inventions, the one snag being the initial cost.

There are also light plastic blinds easily pulled up and down when fixed inside the house, under the roof glass, instead of the usual place on top. One advantage of this arrangement is that the same blinds can be pulled down during frosty weather, and so conserve heat. One of the drawbacks is that they can get in the way of specimen plants, or rather, the plants can be in the way when the blinds are drawn, and so have to be moved or otherwise damaged. Another is that the sun shines through the glass and causes a drier atmosphere, for then ventilators must be opened more fully to allow heat to escape.

By far the most practical method is to fix steel runners up the outside of the roof as shown in plate 21, and fix blinds that let down from the apex to well below the eaves, and also overlap the ends of the house, so excluding sunshine completely when it is wise to do so.

But it cannot be done upon aluminium structures unless there are special hooks or holes made by skilled engineers, such as the manufacturers. It is much easier with wood, for we can screw on lathes or some eyeholes made in steel, and put the blinds just where we wish to.

However, another snag occurs if we affix the blinds right to the top—the ridge ventilators cannot be opened unless the blinds are held well above the glass so that they stand clear. There are firms who erect a framework high enough to overcome this problem, and also supply blinds that can easily be fitted and which can be manipulated without difficulty.

The best way of fitting ordinary blinds is by means of a cross-piece of wood along the ridge. The top edge of the blind is

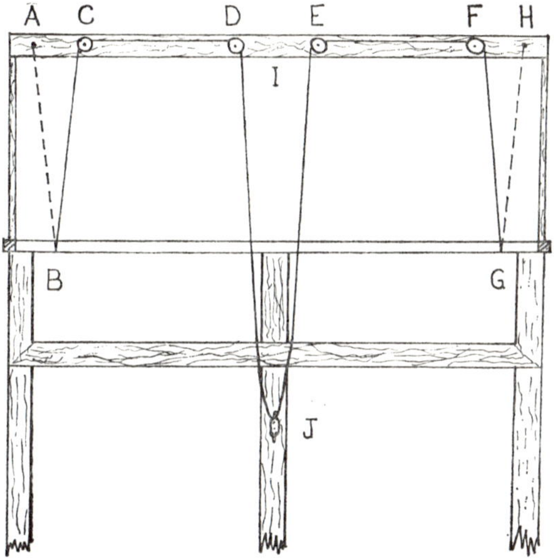

Fig. 14. Fixing cord for roller blind.
Fix end of cord at 'A'.
Bring down under blind to 'B'.
Take cord back over blind to pulley 'C'.
Continue to thread cord through pulleys 'D, E and F'.
Bring down over blind to 'G'.
Take back under blind and fix at 'H'.
Pull slack double cord down at 'I'.
Fix below, to cleat on greenhouse 'J'.

fixed to its underside. Upon its upper side are screwed four brass pulleys through which one continuous cord is threaded. If the roof is 12ft. wide and 6 ft. from ridge to eaves, say, then about 60 ft. of cord is needed. For this area, the blind itself should be 13 ft. wide and 7 ft. down. The cross lath at the top should be 13 ft. long, so overlapping each end by 6 in. The pulleys are fixed to the topside of the lath, or batten, so that there is one at each end, but set in nearer to the middle. The other two pulleys are placed about 6 in. asunder right in the middle (see Fig. 14). The blind is fastened to the underside of the upper lath or batten. And upon the other side of the material, which becomes the bottom, is fastened securely a $2\frac{1}{2}$ in. wide, wooden roller. When that is done the lath is then screwed to the rafters across the top, as near the ridge as practicable. The blind is then resting on the steel runners and the lowered position with the roller hanging down.

We are now ready to fix the cord. One end is attached to an end of the ridge. One way of doing this is to bore a hole in the lath or batten, push the end of the cord through and knot, so that it cannot be pulled out again.

With one end fixed, bring the cord length downwards to the eaves underneath the blind, then carry it back over the top and thread through the first pulley. From there pass it along and thread through the other pulleys in a straight line. When the last pulley is reached, bring the cord down over the top of the blind to the eaves, loop it under the roller and then carry it back to that end of the ridge where it is attached, in precisely the same way as it was at the beginning. There should be plenty of slack cord left. Now use your fingers, or a hook, and pull the cord down the precise middle of the roof where it crosses the two middle pulleys. This should come almost down to the eaves, forming a continuous but double cord, like a pair of reins. Having got it evenly taut, on continuing to pull, the blind will roll up to the top quite easily, and by fixing the loose cord around a cleat, will stay that way as long as you want it to. To let down the blind, merely release the cord. Allowances should be made for possible shrinkages, and similar small snags, and it is best to dismantle the blind and its appurtenances each winter.

There are various materials, the cosiest still being tentage or thick white cotton. Wood laths are often used, as are several forms of plastic material. Sometimes the manufacturers provide special fittings for them. All are worth consideration before a definite choice is made.

CHAPTER 25

Notes upon Heat and Water

SINCE the days when heating the greenhouse was solely through hot water pipes, heated by a furnace within a boiler, many other ways of doing it have been evolved, but none have really proved superior. Nevertheless, they have often been adopted for reasons of expediency and inability to give the time that is necessary for feeding the fires frequently with coal or coke. In this chapter I review most of the various ways and means of heating, and give some addresses where more information about them can be obtained. It is up to the prospective user to consider which is best, or most practical, in the circumstances.

Gardeners like myself are not hot water engineers nor heating experts, but we have to reason and decide which is preferable, in whatever line of equipment is necessary to help us achieve whatever we want to do in the cultivation of any sort of vegetation. To me the chief thing is to know what is available and where further information is obtainable.

Solid fuel boilers need constant attention. The smaller they are, the more often must the fires be stirred and stoked. In large establishments, where there is a range of greenhouses, or two or three large separate structures, the usual form of heating is a very large boiler, generally situated in a stokehole, where the heated water can rise through the pipes and provide a perfect circulation. The fire is often fed with an automatic stoker, to save night attention in very cold weather. Dampers and flues are manipulated by hand or automatic controls, and there must be valve inlets to make good any water shortage, and also small valves at the extreme bends of the pipe system to allow any air that could be in it to be released, so as to avoid air locks.

Particulars of these large boilers and also much smaller solid fuel heating apparatus can be obtained from:

The National Coal Board,
Grosvenor Place,
London SW1

There are smaller boilers that burn solid fuel which have to be built into the wall structure of the greenhouse. They can be attended to from outside. The means of heat transference are 4-in. water pipes that do the job exceedingly well. They are economical to run and, if one is enabled to attend to them as often as is required, there is nothing better.

There are also other small to medium boilers much as above, except that they stand quite free of the structure, the only connection being the pipes that go through the wall. As they are insulated against the cold when in the open, comparatively little heat is wasted. The advantage is that they can be fitted by anyone with ordinary intelligence, and are easy to manage.

Among firms that manufacture one or other of the preceding types of boilers and give much useful information about them are:

H. Phillips Ltd,
King William Street,
Coventry CV1 5NQ

Metallic Construction Company (Derby) Ltd,
Alfreton Road,
Derby DE2 4AQ

Jabez Attwood Ltd,
Eclipsem Works,
Stourbridge

According to one booklet, a portable oil heater is the most economical heater if kept inside the greenhouse. But it can be poisonous to plants. Oil used in boilers instead of coal or coke is next best. Coal, or coke, is slightly dearer than oil. Gas-heated boilers are dearer to run than oil, while the most expensive is electricity.

Oil-burning units are often used today to heat smallish to moderate-sized greenhouses. I favour those that do so by water circulating through 4-in. pipes. These pipes should extend along the front and across the end of a lean-to greenhouse, or in the case of a span-roofed building all around the inside except where the door is positioned. You cannot have too much piping. And the heating unit should be strong enough to keep the pipes quite warm on a frosty night. A gently emitted heat is much better than fierce radiation. We want a nice moistish atmosphere which ample piping is less likely to spoil or reduce very rapidly.

Most oil-burning units are controlled by thermostat. There are two main types. The first is the one that needs no electricity to control the heat or the thermostat. The ones that need electrical aids are wonderfully sensitive and are often installed in large commercial houses.

A popular form of transforming coal boilers into oil-burners is by introducing conversion units. The advantages are eventual economy, once the cost has been cleared, and also the thermostat control, which obviates the need for constant attention. This is a better proposition than removing old boilers and installing modern ones intended for oil heating. *And should oil become scarce or unobtainable, the furnace could easily be reconverted to solid fuel.*

Electricity

Electricity can be a wonderful auxiliary aid in stove and warm greenhouse heating (and in other ways too), but is not ideal as the main heat provider for more than one reason. Firstly, it is too expensive when compared with other heating systems. Secondly, it does not give that lovely calm warmth that is radiated by hot water pipes. Thirdly, it can fail. There are sometimes power cuts when warmth is urgent, and it is possible to lose a whole collection of plants, or at least it may lead to nights of panic.

However, to switch on a fan heater on a very cold night to help the water pipes is surely comforting to the plants, and gives a personal feeling of satisfaction. Heat will not be wasted. The thermostat can be set to the minimum and, if placed near the glass, will switch on or off as dictated by the internal temperature.

There are several models available. The local showrooms of the Electricity Council should have useful booklets available about these and other aids to electrical gardening. Two firms that specialise in these aids are:

Simplex of Cambridge Ltd,
Sawston,
Cambridge

Humex Ltd,
5 High Street,
Byfleet,
Weybridge,
Surrey

Both firms give extensive particulars about their heaters and other electrical or non-electrical aids to greenhouse gardening, and their literature should be thoroughly studied.

Good greenhouse manufacturers also provide suitable auxiliary equipment. It is impossible for me to give any preferences for particular brands, or advise in a completely positive way.

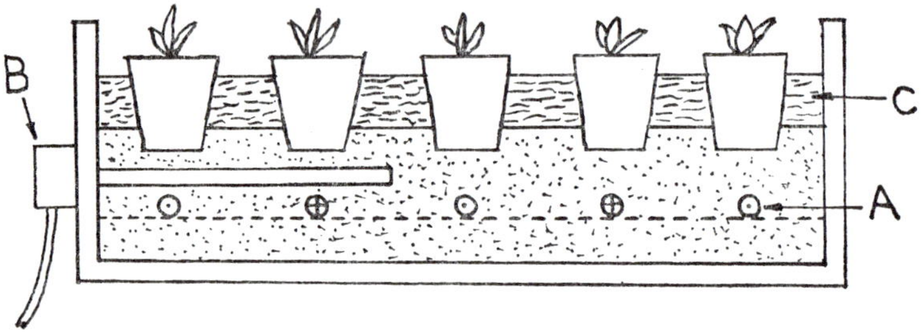

Fig. 15. Soil heating for propagator.
A—Electric cables.
B—Thermostat.
C—Peat.

But I can tell you that these aids should be studied from the long-lasting viewpoint, rather than with optimistic enthusiasm concerning their novel usefulness. I reiterate that they are not as good as first-class human management, but they are better than poor or indifferent care. And they are most useful upon special occasions, such as during holidays, or when other business is more important.

Electricity can also be used in providing the additional heating conditions written about in the first two chapters. In Fig. 4, there was given an illustration of how to give 'bottom heat' to the propagator by greenhouse pipes. Instead of this, it is possibly better if, instead of water pipes, the propagator bed is fitted out with electrically heated cables (see Fig. 15). Several firms, including those already mentioned, provide them, together with all the necessary thermostatic and automatic controls. They also provide proper propagating cases to hold the floor materials and the receptacles for the cuttings.

Mist Propagation

This is one of the fashionable pastimes amongst keen amateurs, as well as being the usual method used by nurserymen of increasing stocks of trees, shrubs and plants. Naturally, it is a matter of degree. The large nurserymen will have considerable equipment, while the limited amateur will be content with and proud of his single jet that helps him to root his few cuttings of whatever is fancied.

In order to have a small installation it is necessary to have bottom heat, supplied either by warm pipes or by underground cables. He must have a mains water supply with sufficient pressure to mistify the moisture as it is forced through the jet. He must have an electric solenoid valve that turns on, and shuts off, the water automatically when it gets the message from the part called the 'leaf' when it is dried by the atmosphere.

A first-class propagator consists of the underground heating cables that are controlled by thermostat, and a really good mist jet. In the summer when the propagator is not filled with cuttings, the cable-heating equipment can be switched off, but the mist jets should be retained to help create the desirable moist atmosphere throughout the house. Other jets can also be put elsewhere in the house if deemed desirable, although this should be carried out cautiously.

Hot Air

Although not yet generally practicable in greenhouses of modest dimensions, there are now several forms of heating by air involving the use of coal, gas or oil, together with electrical regulators that are both most efficient and economical. The oil and gas heaters are generally fed from a tank or mains, and by the intensive working of the machinery, including fans, the hot air is blasted to wherever it is wanted. In some instances the machine is inside the large glasshouse. In others, the heat is transferred to where it is needed through polythene tunnels. When these are laid along the floor of the house, the heat is released from them through small apertures, much as water is, through perforated hoses in the garden.

Not only do these machines produce the heat and project it where needed, they also control ventilation according to the amount of sunshine, regardless of temperature, although that, too, is controlled and the heat automatically shut off. Another aspect is moisture control. The machines are described by the various manufacturers—of which there are several in this country and abroad—in their own jargon, as having such features as: 'automatic ignition'; 'gear type fuel pump'; 'triple skin heat exchanger'; 'optimum humidity'; 'two-wire unpolarised interconnection', etc.

While it is possible that someone may read this book who is anxious to know more about these developments, the majority of us are aware that they are outside the horizon of the amateur plantsman. But they do exist, and if anyone cares to investigate, a lead can be obtained from the Ministry of Agriculture, or a nursery trade journal.

There are also three types of hot air projectors for heating small greenhouses. One is powered by paraffin, another by gas, and the third by electricity. In the case of paraffin, the lamps with incandescent burners are useful to those who want merely to keep frosts out, but if they go wrong there is trouble through fumes and muck. In the case of natural gas, of recent distribution, there are claims that internal heating improves the atmosphere through emitting damp CO_2. It is not for a gardener to argue with science, and I prefer to let all those who have faith in natural gas to use it. I fear that any stove or warm greenhouse heated exclusively by an internal gas stove is not a good place for human beings, nor permanent ornamental plants. Notwithstanding, I realise that this may be prejudice, and if anyone has the normal mains supply it could be a simple and easy way to obtain good heat. Any Gas Board office or showroom, will provide the literature.

Actually, in addition to the incandescent paraffin lamps, there are now thermostatically controlled heating systems that burn that form of oil. Instead of wicks that draw the paraffin upwards, the system is one where pipes feed burners below the tank level. This eliminates the danger of draughts causing the lamps to flare up and emit soot and, as the thermostatic valve controls the burner, there is little to go wrong. The manufacturers claim the same qualities for these that the gas men do, i.e. that CO_2 is produced, as well as moisture, both being conducive to plant life. A firm that designed and makes such a heater is:

Cooper-Walker Engineering Ltd,
Dunbar,
Scotland

A possible drawback of both the paraffin and the natural gas heaters is that it is impracticable to have a special propagator in the greenhouse unless there is also electricity, or the normal hot water pipes heated by a proper boiler. This should be taken into account when installing such things. No doubt something could be invented, if thought were given to the problem. And there may be propagators that can be heated by these means, but I have yet to find them.

Conclusions

To sum up the problems of heating greenhouses, so much depends upon so much, when making the choice of equipment. For instance, if the new acquisition is a lean-to or conservatory adjacent to the residence, and the latter is already centrally heated from a boiler (regardless of what it burns) by a system of radiators, it would be most practical to extend the heat to the plants by putting in additional radiators, but only if the boiler was sufficiently large to carry the load. If it could do this moderately well, then it would be a good proposition to instal portable fan heaters with thermostat controls as well, to boost the temperature on a cold night. In such circumstances the outlay and running cost would not be uneconomical.

Should the new plant house be of moderate proportions and stand apart from the residence, it would be well to instal one of the small outside boilers, either of solid fuel if regular attention can be given, or by thermostat if there is an oil feed from a storage tank. The boiler would keep the greenhouse heated by the pipes in the best manner possible. But also, if an electricity supply can be introduced with suitable power points, this would be most useful to boost temperatures on cold occasions, and also be helpful in many other ways, such as controlling moisture, heating soil and beds and giving light and ventilation.

Regarding water, while rain collected from the roof and stored in an interior tank is quite a good proposition, it is not as essential as is a mains supply. For the stove or warm greenhouse, moisture is just as important as warmth. While there are many ways of warming the house, there is only one practical means of obtaining a regular supply of the essential water, and that is by a pressure tap from the mains.

CHAPTER 26

Notes upon some Useful Auxiliary Aids

VENTILATION and its relation to air movement has been discussed several times in previous pages. The advantage of a portable fan-heater as a means of supplementing the standard heating system during a cold period has been emphasised, and also its value in stimulating air movement by the fan alone when the heat has been switched off at the instruction of the thermostat. There is no doubt whatever that such air movement is good for plant cultivation. It spreads a lovely even temperature and also discourages mould and rots from attacking the plants, which they sometimes do when there is stagnation. But it is often disconcertingly noisy.

Fans

Where there is a good heat promoted by the permanent system, especially when warmth is transferred through water pipes, I consider a better proposition than the additional fan-heater is to have one small, independent fan working silently in the house all the time so that there is a perpetual slow and gentle circulation of air both by night and day, and in winter and

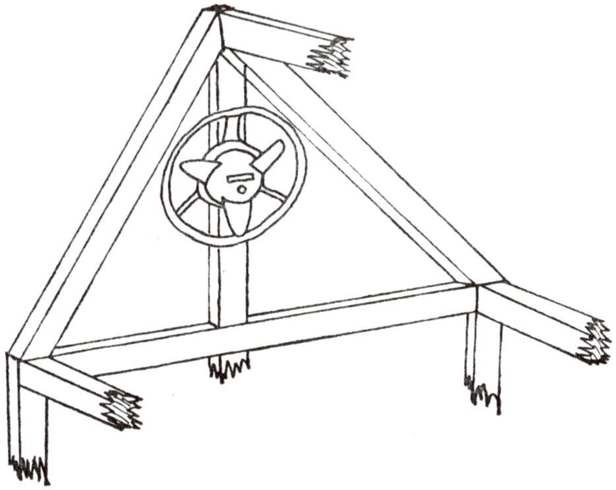

Fig. 16. Small fan.

summer, regardless of the temperature or the heating apparatus. The two firms already mentioned, Messrs Simplex and Messrs Humex, list models of the type I mean.

Should there be a severe wintry spell when additional heating is advisable, the ordinary household electric convex heater, or any other heat-giving apparatus, can be used, as long as it is of a type that will not emit dangerous fumes, whether or not it can go wrong. There are not many nights of any year when cold is so bad (see Fig. 16).

Ventilators

The same small fan will also help to circulate the air in spring and summer when the ventilators are open, according to necessity. Not only will the air be exchanged, but it will also be stimulated out of and into the house, so making it less essential to have extensive openings. We need not worry about the size of ventilators half as much when there is no fan.

In summer the glass is shaded when the sun is at its strongest, and this is another reason why extensive ventilators are less imperative than they are in cooler, yet unshaded, houses that demand lower, drier temperatures. Still another cooling element is the damp atmosphere, which must always be provided for tropical plants.

However, there are times when ventilation is necessary, as described in the first two chapters, and there are automatic, self-operating controls that will open and shut ridge sashes according to the temperature. They need no driving power, such as electricity, but rely upon the principle of expansion and retraction of an element within a sleeve.

As I wrote earlier, these gadgets cannot anticipate. But they are acceptable for those who cannot be in attendance all the time. It is nice to know that the gadget will open one of the ventilators upon a bright early summer morning whether you are there or not. It will also close down the ventilator again when the sun wanes. When such a control is fitted it can be allowed to do its work through spring, summer and autumn, but may either be dismantled or made inactive during winter.

The other form of ventilation that is frequently advertised and used is the extractor fan, inserted in the glass at one end of the greenhouse, which changes the air automatically. While

I have seen it working well in cool greenhouses, I feel reluctant about fitting it to any place where a good temperature is imperative. Too much air drawn in during cold weather necessitates correction by increasing artificial heat. In summer, the normal methods of ventilation, together with shading and damping down, are enough.

One of the routine tasks old establishments insisted upon was keeping the spaces between the panes of glass, where they overlap each other, quite clear of pollution and fungi by pushing a flat piece of tin or metal down between them. A flat metal label conveniently bent so that one end serves as a handle does quite well. Not only does this enable an ample change of air to take place in winter but it also improves light, and discourages pests from making their homes there.

Automatic Watering

There are various devices aimed at reducing to a minimum the chore of watering plants. One of these is to put a special layer of sand upon the bench and keep it moist by a continuous, slow but sure, supply of water from a controlled feed. When the bottoms of the pots or receptacles are slightly buried in the sand, capillary attraction draws moisture from the sand to the roots through the drainage holes.

Whether or not this labour-saving method of keeping pot soils moist is successful depends upon such factors as the flatness of the bench, the size of the pots containing the roots of the plants, and whether the compost is of the right texture to draw the moisture. Care has to be taken to ensure that drainage crocks do not stop close contact with the floor. Uncrocked plastic pots with their several holes in the base are better for this than are clay pots, that generally have only one water escape, which must now be in reverse.

I, personally, do not like slush beds, for more than one reason. Chief of these is appearance. They soon get messy and, if pots are moved, there is an unmistakable mark where they stood. Furthermore, the pots containing ornamental plants must vary in size according to the age or species of the occupants. And there are changing seasons when the various plants need comparatively much or little water. These capillary beds are quite good for batches of plants that are all the same size and need the same degree of moisture. But we are growing ornamental plants, and they should always be given surroundings that show them off well. Furthermore, for staging them it is necessary to stagger them by placing some on stands or upturned empty pots. Capillary watering is very utilitarian and, accepted thus, can save a deal of work.

There are also other new forms of watering continuously being evolved, although they are based on the system of upward flow, instead of going down to the roots. There are felts and patent materials being employed to try to stop the slush effect.

The alternative principle of watering, by feeding into the top of the pot instead of the bottom, is by a pipe with a nipple at the end. This is being done in large commercial nurseries where pot plants are grown by the thousand. It involves just as many pipes leading from the water supply, so that each gets the amount of water considered right for its welfare. The spout, or nozzle is adjustable to make that possible.

The other system is to have nozzles or spouts inserted in one long pipe at regular intervals, so that each is placed over one of the pots. The water is fed through a cistern on a principle much like that of the lavatory flush. When the cistern becomes filled the water flows down the pipe until all is gone; then the cistern slowly fills up again. The nozzles or spouts are mostly adjustable so that, as the water flows along the pipe, each plant gets only the amount that is considered requisite.

My chief objection to these automatic watering gadgets is again appearance: to see pipes and nozzles decorating the staging and the pots irritates me. Ornamental plants in leisure greenhouses are grown for beauty, and all around them should be in harmony. Any aids must be judged as much for their looks as for their usefulness. On the other hand, it has to be conceded that, for the casual gardener who cannot be on hand all the time, these watering systems are a boon. It is for that reason they are brought to notice.

Patent Pots

Another form of easing regular watering is by means of pots and containers that provide their own water reservoirs in the form of built-in chambers, that only need to be filled weekly, or at whatever interval of time is required. During the years there have been many patented, but they have either been too expensive, or have not passed the test of time.

However, Simplex of Cambridge have now evolved a system whereby troughs are inter-connected, and no matter whether they are in flat alignment or are tiered, there is a connecting pipe that will keep the roots supplied with just the right amount of water. While not altogether practicable in the warm greenhouse, the idea is not impossible, and is worth consideration.

Air Humidifiers

Air humidifiers are being much used in private homes, offices and shops, where heating systems create dryness in the air. In most cases, water is vapourised into the atmosphere by dispensers controlled by electricity. The chief drawback for practical use in keeping warm greenhouses humid is that they are unlikely to fill the air with enough moisture, and they must be continually replenished, unless there is a controlled ball-cock supply of water. These problems are, no doubt, being overcome. One supplier of several models is:

Air Improvement Services,
21 Napier Road,
Bromley NR2 9JA

Sterilisers

If a reliable supply of loam or potting soil is obtainable, this is one way of ensuring that the compost mixture used for plants is of standard quality. Also, one is sure of adding the right sort of peat and sand. There are several excellent models that can be controlled easily without fear of over-doing or under-heating the contents. Some are fitted with a synchronised timer control unit that switches off the heat when the soil has been baked enough. This is for those who want to sterilise a bushel of soil at a time.

There is also an electrically controlled steriliser that burns paraffin and works on an endless chain principle, the soil being placed in a trough that passes through the heated chamber much like bread does in a baker's oven. This is an ingenious machine, more helpful to commercial and municipal gardeners than to those who have only one small greenhouse.

Insecticide Vaporisers

One way of being completely free of insect pests is to instal an insecticide vaporiser that is controlled by electricity. By putting in different canisters according to the pest, such as red spider mite or whitefly, and keeping it going at minimal expense day and night, there is complete eradication, whereas if only instant spraying or fumigation is carried out, there is almost sure to be a recurrence of a particular trouble. But whitefly and red spider should never be bothersome in a warm plant house. However, if they or other insects or fungal bothers are persistent, and one is happy about the use of the poisons the canisters contain, there is no surer way of eradication.

The address of a firm that hires these machines out, supplies the insecticides and fungicides in their canisters, and runs a maintenance service is:

Shepherd's Aerosols Ltd,
Shernfold Park,
Frant,
Tunbridge Wells,
Kent

Although I have used their dispensers for several years in large greenhouses, I am rather reluctant to do so in the tropical house. While some cannot detect the smell of the insecticides permeating the atmosphere, I have never failed to do so, and to live with it all the time is not conducive to good health.

Alternatively, there are numerous pesticides that can be used in many forms, such as smoke generators, sulphur candles, sprays, aerosols, sponging washes, dusts, baits and strips. The last-named can be hung in the greenhouse, and are useful for dealing with insects over a limited period.

A firm that stocks a very wide range is:

Joseph Bentley Ltd,
Barrow-on-Humber,
Lincolnshire

The same firm offers a vast range of gardening aids that are useful to the owners of warm greenhouses. They include John Innes composts, peats, grit, loam, osmunda fibre, soft soap, liquid carbolic soap for washing and disinfecting staging and greenhouse interiors, syringes, sprayers, netting material, raffia, labels, pots of every description and size, wire baskets, and many other greenhouse accessories.

Appendix

Names and Addresses of Plantsmen, Seedsmen and Specialist Societies

IN the Introduction a brief allusion was made to the collecting of plants for the greenhouse. Local florists were mentioned and the studying of seed catalogues was suggested. A fitting conclusion to this work is to give addresses of firms and organisations that can help the reader in such a quest.

It is impossible to vouchsafe for the accuracy of all the addresses. There is constant change everywhere. The plants themselves have their titles changed from time to time, and it is almost impossible to keep abreast with this aspect of nomenclature. Nevertheless, it is hoped that the information will be helpful, and will be accepted in the spirit given.

House-plants, including poinsettias

Thomas Rochford and Sons Ltd,
Turnford Nurseries,
Turnford,
Nr Broxbourne, Herts.

These people are wholesale only, but supply very many retail shops.

House-plants, including saintpaulias

Elm Gardens Nurseries,
Church Road,
Claygate,
Surrey

Berkeley Nurseries,
Dudsett Lane,
Cranford,
Hounslow,
Middlesex

Indoor Plants, including bromeliads

Holly Gate Nurseries Ltd,
Billingshurst Lane,
Ashington, Sussex

Indoor Plants, including begonia species and bromeliads

Mr B. Wall,
4, Selbourne Close,
New Haw,
Weybridge,
Surrey

Greenhouse and Conservatory Plants

Longmans Ltd,
154 Fenchurch Street,
London EC4.

Greenhouse Plants, including gardenias and hibiscus

Shopwhyke Nurseries,
Chichester,
Sussex

Orchids

Black and Flory,
The Orchid Nursery,
Nursery Lane,
Middle Green,
Langley, Nr Slough,
Bucks

Burnham Nurseries Ltd,
Kingsteignton,
Newton Abbot,
Devon

Seeds

Major V. F. Howell,
Fire Thorn,
Oxshott Way,
Cobham,
Surrey

Thompson and Morgan Ltd,
Ipswich,
Suffolk

Samuel Dobie and Son Ltd,
Chester,
Cheshire

Note. Most seed lists contain some names of plants that need a warm greenhouse.

There are at least four specialist societies devoted to plants we grow in warm greenhouses:

The British Bromeliad Society

The Orchid Society of Great Britain

National Begonia Society

Saintpaulia and House Plant Society

While societies like these are very helpful, and are worth pursuing, if one is interested in such plants, the addresses of the officials frequently change, and the best way to make contact is through one of the firms who deal in the respective sorts of plants. Or, if that is too difficult, it is most probable that the Society of societies can help. The address is:

The Secretary,
Royal Horticultural Society,
Vincent Square,
London SW1P 2PE

Index

(Family names are given in small capitals)